UNDERSTANDING IMPERILED EARTH

UNDERSTANDING IMPERILED EARTH

How Archaeology and Human History Inform a Sustainable Future

TODD J. BRAJE

Smithsonian Books

WASHINGTON, DC

Published by Smithsonian Books
Director: Carolyn Gleason
Senior Editor: Jaime Schwender
Editor: Julie Huggins

Edited by Erika Bŭky
Designed by Gary Tooth

This book may be purchased for educational, business, or sales
promotional use. For information, please write: Special Markets
Department, Smithsonian Books, PO Box 37012, MRC 513,
Washington, DC 20013

Library of Congress Cataloging-in-Publication Data

Names: Braje, Todd J., 1976– author.
Title: Understanding imperiled earth : how archaeology
    and human history inform a sustainable future / Todd J. Braje.
Other titles: How archaeology and human history inform
    a sustainable future
Description: Washington, DC : Smithsonian Books, [2024] |
    Includes bibliographical references and index.
Identifiers: LCCN 2023031727 (print) | LCCN 2023031728 (ebook) |
    ISBN 9781588347596 (hardcover) | ISBN 9781588347602 (ebk)
Subjects: LCSH: Environmental archaeology. | Human ecology—
    History. | Sustainability.
Classification: LCC CC81 .B73 2024 (print) | LCC CC81 (ebook) |
    DDC 930.1028—dc23/eng/20230719
LC record available at https://lccn.loc.gov/2023031727
LC ebook record available at https://lccn.loc.gov/2023031728

Printed in USA, not at government expense
28 27 26 25 24   1 2 3 4 5

To my grandfather, Bill Sampson. Papaw has faithfully read
every one of my books. Each time I sat down to write
this one, I thought of him, the questions he might ask, and
what he might want to know. I hope it was a good chat.

●

CONTENTS

Chapter 1

NATURE, HUMANS, AND HISTORY

For nearly all of the past fifteen academic years, I have had the good fortune to teach my favorite undergraduate anthropology class, Introduction to Archaeology and World Prehistory. As a first-year college student, I took a similar course, Buried Cities and Lost Tribes, which helped spark my interest in the field. The snappy title was just a way to hook students thumbing through a mountain of class options in their course catalog: in reality, it offered a survey of world prehistory, an overview of the physical and cultural evolution of humans, and an introduction to archaeology.

The course I teach today, much like the one I fell in love with thirty years ago, reviews the astonishing journey of humans, from their evolutionary origins in Africa to their initial migrations around the world, the rise of agricultural societies, urban states, writing systems, and the many other hallmarks of human society. The story covers some seven million years. For me, there is no tale more exciting and mystifying than that of how and why human societies have come to be what they are today.

Most of my students tend to be freshmen overwhelmed by their new routines and lives on campus, so I try to capture their attention immediately. Without any introduction or commentary, I begin the class by screening the first few minutes of *Raiders of the Lost Ark*. The opening scene depicts the ruggedly handsome American archaeologist Indiana Jones entering a Peruvian temple deep in the jungle in 1936. To gain entry, our hero must navigate deadly hazards: a mass of tarantulas, a pit that he swings across by unleashing his whip, and a booby-trapped floor. He manages to remove a golden idol while narrowly avoiding being crushed by a giant boulder. After escaping the temple, he is cornered by an evil rival archaeologist who steals the idol but inadvertently allows Indy to escape aboard a waiting seaplane.

Of course, the movie, and its depiction of archaeology and archaeologists, is Hollywood fantasy, but it has probably done more than anything or anyone to boost the mystique and popular reputation of the field. I use the clip to discuss what archaeology is not and should not be. Indy is not an archaeologist; he is nothing more than a thief. But movies like the Indiana Jones series, along with popular television shows, magazine articles, and news stories about exciting archaeological discoveries, bring students flocking to my introductory course. Like many people, they view archaeology as a form of intellectual entertainment. I try to leverage this interest and enthusiasm and teach my students not only how ethical, scientific archaeologists really study the past but also why it is important that we continue to do so.

The Spanish philosopher George Santayana famously wrote, "Those who cannot remember the past are condemned to repeat it." Variations of this quote appear on countless websites, on inspirational posters with dramatic landscape backgrounds, and in the halls of academic institutions. They are not wrong. But there is more. The past is inescapable. All of our decisions and those made by our ancient ancestors have shaped the world we live in. Understanding these decisions, identifying the incremental changes that have altered our planet, and engaging with deep history helps us imagine the world we want to inhabit and the one we hope to leave for our children and grandchildren.

My graduate education in archaeology began at the University of Florida, where I completed a master's degree in anthropology with a specialization in archaeology, studying what was then the oldest known pottery in North America. In the Laboratory of Southeastern Archaeology, I sorted through bags of crumbling pottery fragments (*sherds* to archaeologists) excavated from three archaeological sites on Stallings Island, South Carolina, that were around four thousand years old. For two years I peered through a microscope at their edges, classifying how they were made and the materials they were made from. I meticulously described their surface decorations and hypothesized about what these fragments could tell us about ancient social relationships.

I began my master's degree immediately after returning from the Kingdom of Tonga, a small island nation in the South Pacific, where I served as a Peace Corps volunteer. I was becoming increasingly concerned about the environmental challenges facing our world—including global warming, pollution, plant and animal extinctions, overfishing, deforestation, and rising sea levels. My time in Tonga gave me a greater appreciation of the risks these changes posed for communities living on islands and along coastlines. During my master's research, I realized that the microscopic investigation of ancient pottery wasn't the type of archaeology that excited me. I knew that archaeology could speak to the environmental challenges I had witnessed in the Peace Corps, heard about on television, and read about in the newspaper.

Before we go any further, let's clear one thing up. By *history*, I am referring not only to the approximately five thousand years of recorded human history since the advent of writing systems but also to the geology and biology of planet Earth: mountains forced upward by colliding continental plates, the fossil remains of plants and animals large and small, and the soils and sediments that detail local environmental conditions. History is recorded not only on clay tablets or book pages but also in tree rings that chronicle ancient climates and pollen grains buried in layers of lake-bottom sediments. History is found in the things people left behind over thousands of years, the tools they made, and the foods they consumed or tossed away. History is told in rock art, etchings, and ancient graffiti in caves and rock shelters. History is written into our very bodies, into our DNA, into traumas undergone and healed, and into our bones and teeth, which are shaped by the foods we eat and the water we drink. History is imbued in the languages we speak, the sounds and words we utter, and the ways we invest these with meaning. History is passed from one generation to another in the stories we tell—real and imagined—and the cultural traditions we carry on.

History is also written into the ruins and remains of bygone civilizations. People of the past, just like us, confronted obstacles big and small; they faced decisions about how to structure their lives; they interacted with the world around them; and they dealt with the challenges posed by nature or of their own making. In general, leaders of social groups surely made the best decisions they could. Some allowed people to prosper, and others had disastrous consequences. Although these decisions may have been made when human societies were smaller, less interconnected, and operating on more localized scales, they shaped the world around them and left an enduring legacy on the planet. We live with the consequences of those choices, just as our children and great-grandchildren must live with the consequences of the choices we make.

This book is about this critical importance of the past. It is not, however, about the lessons that can be learned from the suc-

cesses and failures of ancient civilizations or cultures. That topic
has been well covered. Jared Diamond's *New York Times* bestseller
Collapse: How Societies Choose to Fail or Succeed is probably the
best-known example. A more recent example is Annalee Newitz's
Four Lost Cities: A Secret History of the Urban Age. My book takes
a different approach by exploring the ancient connections between
humans and the environment and how humans came to exert such
tremendous pressure on Earth's ecosystems. Without understanding the extent of forests and their composition eight thousand years
ago in Europe, prior to widespread land clearance by early agriculturalists, how can we hope to rebuild natural ecosystems across the
continent? Without understanding historical fluctuations in greenhouse gas levels and temperatures, how can we set modern targets?
If we do not know how many fish and sea mammals were in our
oceans before massive commercial fishing and hunting began, how
can we know what a healthy ocean looks like? Without the past and
perspectives from history, we are planning a path forward without
consulting a roadmap.

At times, the complex links between past and present can be
difficult to understand because humans have the tendency to think
in dichotomies: good/bad, light/dark, healthy/unhealthy. Binary
thinking provides a sense of certainty and comfort, especially during uncertain or troubling times. By enabling us to make rapid judg-
ments, in some ways it also helps keep us safe. But such categorical
thinking leaves little room for understanding nuance and gray areas.
So when we think about environmental challenges, it is easy to fall
into binary thinking: either something is natural or it is cultural.
If we believe that "natural" places are those untouched by human
hands, the solution to our environmental problems may seem to be
simply a matter of removing human influence. But this nature/cul-
ture dichotomy oversimplifies a very complex relationship. Humans

and nature are inextricably linked. Anthropologists sometimes call this connection an *entanglement*: that is, influence works both ways. For millennia, humans have been altered both culturally and biologically by our interactions with the natural world, while at the same time we have altered the natural world to suit our needs.

Take, for example, the transition from primarily hunting and gathering to agricultural societies. Approximately ten thousand years ago (in some parts of the world a bit earlier and in other parts a bit later), humans domesticated plants and animals (see chapter 2) and launched the so-called Agricultural Revolution. Domestication of animals and plants was possible only through human selection of desirable behaviors and traits (such as docile dispositions in live-stock and large seeds in plants) that changed their genetic makeup. Once domesticated, these plants and animals could no longer survive in the wild, and humans had to transform their own way of life in order to support these altered species: sowing seeds, preparing fields, storing provisions, selectively breeding animals, and settling in permanent locations. The transition to farming brought about the rise of city-states, centralized leadership, complex sociopolitical systems, and, eventually, modern human societies. It also altered our DNA. Scientists have discovered specific genes that changed in humans during and after the transition from hunter-gatherers to agriculturalists. These include variants on or near genes associated with height, the digestion of lactose in adulthood, fatty-acid metabolism, and vitamin D levels.

The bottom line is that there is no "natural" world, if we conceive of *natural* as meaning untouched by humans. Likewise, there is no cultural world without nature. It's better, then, to think of nature/culture not as a dichotomy but as a continuum. Some places are more natural and less culturally shaped than others, just as others are more culturally shaped than natural. Even the most isolated places on Earth have been altered by human action, from the highest mountain peaks—which now have less annual snowpack because of anthropogenic climate change—to the most isolated islands of the Pacific Ocean, which have microplastics embedded in their sandy beaches. The nostalgic notion of pristine places on Earth, untouched by human activity, has been influential in discussions of environmental conservation, but their existence is an ecological fantasy. There is no way to turn back the clock to a time before human activity altered our planet. The natural world is always and has always been changing.

THE NATURE OF CHANGE

Earth and its ecosystems are in a constant state of flux. Daily, seasonally, yearly, and over human generations, change happens at a small scale. The Sun rises in the east, and as the Earth turns, it appears to travel across the sky; plants and animals respond to its daily cycle. Ocean tides ebb and flow as water is moved by the gravitational pull of the Moon; beaches and rocky shores are covered or exposed. The tilt of the Earth's rotational axis, either away from or toward the Sun, creates distinctive seasons. Plants bloom, fish spawn, and birds migrate in response to these changes. The human measures of days, weeks, months, and years are built around these cycles and structure our lives and cultural traditions.

Beyond the scale of a human lifetime, much more dramatic environmental changes have taken place. Over the last 2.6 million years, changes in solar insolation (the intensity of heat from the Sun reaching Earth) have resulted in dramatic shifts in temperature. Seven cycles of glacial advance and retreat (ice ages) have occurred over the last 650,000 years. During the colder periods, global average temperatures dropped some 6°C (11°F). Glacial ice expanded in polar and high-altitude regions, and sea levels dropped by about 120 meters (400 feet), as ocean water evaporated, fell as snow, and accumulated in massive glaciers and ice sheets.

In the early twentieth century, the Serbian mathematician and scientist Milutin Milanković spent decades investigating the causes of these long-term, global, and cyclical climatic oscillations, ultimately demonstrating that they were related to changes in the position of the Earth relative to the Sun. He used geometry, celestial mechanics, and theoretical physics to demonstrate that these shifts were due to three factors: changes in the wobble of the Earth's daily spin; changes in the tilt of the Earth on its axis; and changes in the Earth's yearly path around the Sun (the precession of equinoxes). The resulting climatic variations led to ecological instabilities, and in some cases to the extinction or extirpation (local removal) of keystone plant and animal species, those species that have a disproportionately large influence in shaping ecosystems. Scientists have documented wholesale landscape transitions from grasslands to forests or vice versa as the result of such ecological shifts.

When we extend our temporal lens back into deeper geologic history, the changes are even more dramatic. Millions of years ago, the Earth was a vastly different place. Continental plates have slowly drifted to their present locations, building the Himalaya, the Andes, the Alps, and other mountain ranges as they pushed against one another over the past twenty million years. South America and Tasmania drifted apart from Antarctica 35.7 million years ago. This separation enhanced ocean circulation, which in turn decreased carbon dioxide levels, creating a global cooling trend.

In just a few thousand years of human existence, our species has effected changes to the planet that rival those of geologic forces. To someone standing on a sidewalk in a bustling city as people negotiate their way to cubicles in high-rise office buildings, or sitting

An overview of modern-day Mexico City and the Valley of Mexico, showing vast urban sprawl broken up by pockets of green space. The mountains that ring the city are obscured by a thick layer of looming smog. (Image via Wikimedia Commons.)

in a car stuck in a traffic jam on a freeway connecting suburbia to an urban center, the world seems entirely constructed by humans. Years ago I flew from San Diego into Mexico City. Situated in the Valley of Mexico, a high plateau, greater Mexico City has a population of over twenty-one million people, making it the most populous city in North America and the sixth largest metropolitan area in the world. From my airplane seat I gazed in awe at the scale of the cityscape. Ringed by beautiful mountain peaks, the entire valley was a tangled, sprawling labyrinth. Green spaces and lakes occasionally broke up the urban jungle of homes, buildings, and roads, all enveloped in a brown haze of smog. The view highlighted for me the magnitude of the human alteration of our planet.

Many of us seek escape from this reality in the seemingly pristine places that remain: hiking trails in national parks, mountain

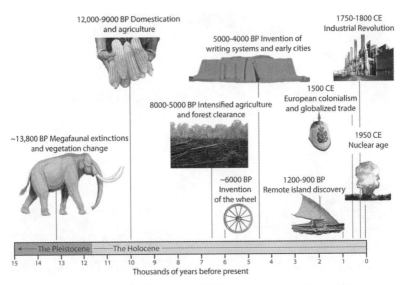

Timeline of major global transitions over the last fifteen thousand years of Earth's history. (Images via Wikimedia Commons.)

peaks, rainforests, and wildlands. These places offer solace and escape from our busy lives and the trappings of modernity. They act as refugia for plants and animals—havens less affected by the changes in surrounding areas—and, in some cases, as biodiversity hotspots that shelter a rich variety of plants and animals. Yet even these places have been influenced and altered by humans.

Humans have been manipulating the Earth's environment for such a long time and in so many ways that human-induced alternations have come to seem like part of the natural world, the way things have always been. Where I live in San Diego, parks, roadways, and open spaces are often shaded by massive eucalyptus trees. While this species may seem well adapted to deal with the aridity of southern California, eucalyptus is a recent import to North America, introduced from Australia in the 1850s during the California

10

gold rush to provide lumber, fuel, and windbreaks. Another example is tumbleweed, made famous by Hollywood Westerns. No shootout would be complete without tumbleweeds blowing across hardpacked dirt roads as bullets whiz past the protagonist's five-gallon hat and shiny sheriff's badge. But this iconic plant is actually a Russian thistle, accidentally introduced as part of a flax seed shipment to the American West in the late nineteenth century. Even the dandelions and crabgrass that invade the classic US suburban lawn are plants native to Europe.

All of these are relatively recent translocations and introductions, but humans have been moving plants and animals from one location to another for thousands of years. The point is, we tend to have a short memory when it comes to our surroundings. Within a generation or two, an introduced species or an ecological shift becomes normalized. We tend to evaluate the state and health of our environments by the yardstick of our own experiences. Our perception of the world when we were children or young adults, often influenced by media images, becomes the standard by which we measure all subsequent ecological changes. Our memories of what the world used to be—real or imagined, exaggerated or not—often become our reference points. We recall images of the ocean filled with large sport fish like tuna, swordfish, and giant groupers; the wild African savanna teeming with giraffes, gazelles, and other wildlife; the prairies of the North American heartland filled with herds of bison; and (though no one alive can now remember them) skies darkened with flocks of migrating passenger pigeons.

For environmental scientists, *baselines* are reference points for evaluating the state and health of ecosystems. They are the standards by which we measure change. To build effective environmental conservation plans, we must set our ecological baselines at references points prior to intensive human impacts, such as the rise of the commercial fishing industry that has depleted the oceans of its bounty; the illegal poaching and encroaching agricultural and industrial economies that have driven many iconic African species to the brink of extinction; and the overhunting of bison and passenger pigeons for skins, cheap meat, and sport. Although we strive to set our baselines at realistic starting points, at conditions that we believe to be natural or desirable, we tend to forget the changes that human activity has wrought, and we accept an increasingly degraded environment as natural. This "shifting baselines syndrome" has become a revolutionary concept in environmental conservation, one that situates history as crucial for addressing our environmental crisis.

BASELINE SHIFTS AND HISTORICAL REMEDIES

There are two basic categories of baseline shifts: personal and generational. These shifts are best thought of as types of amnesia, because we lose knowledge without even being aware that it is happening. Personal amnesia involves a failure to remember how something used to be during your lifetime. For example, you might not remember that during your childhood explorations of your backyard, birds, lizards, and frogs were common sights. You forget how the world once looked, even places with which you are intimately familiar. Generational amnesia involves a loss of information from one generation to the next. A local fishing hole that you once considered pristine is still viewed as unspoiled by your children, despite the loss of biodiversity, the introduction of invasive species, and other degradation. The fish you used to reel in from that honey hole were larger, more abundant, and of a different species than those captured by your children today, but they are thrilled by the smaller fry on their hooks.

Baseline shifts in our everyday lives illustrate how they happen in the natural world. The gradual protraction of your morning commute over months or years can be imperceptible, with just a few more commuters, a few more cars, and a few more seconds or minutes added to your drive every few months. Without your realizing it, a "normal" commute has stretched to forty-five minutes, and you don't quite remember how long it used to take. Your personal baseline has shifted.

Scientists and their research can also fall victim to shifting baselines. Many scientists tend to accept observations they made in their early field and laboratory work as the baseline for evaluating subsequent changes. Changes in forest cover, animal populations, and species compositions, for example, are viewed in a positive or negative light based on a researcher's early career findings.

The scientific process can also be affected by generational amnesia. Data collected and interpreted by previous scientists are often viewed as deficient by contemporary leaders in the field. The way in which scientists conducted fish counts prior to modern electronic instrumentation might be viewed as inadequate and potentially flawed today. When new modeling techniques are developed, more rigorous statistical analyses are applied, and new methods for collecting data become the standard, the scientific perspectives of the past can be brushed aside. The result is a gradual shift in baselines and the creation of inappropriate reference points for evaluating the health and structure of ecosystems. Our perception of what is natural and normal shifts based on an increasingly degraded standard. When environmental scientists fall victim to personal and generational amnesia, their conservation and restoration plans may aim to restore a degraded system to one that is only slightly less degraded. When the public falls victim to historical amnesia, it becomes very difficult for conservation scientists and advocates to

convince people that the environment is degraded and that something needs to be done about it. If you do not remember a pristine
environment or recognize the degradation that has occurred, why
would you spend time, effort, and money fixing the problem?

One way to combat our historical amnesia is to look for control sites, or places that have been protected from, or less affected
by, human action. There are very few such places left on the planet.
High-altitude environments and inhospitable deserts offer some
insights, but the environments most in need of restoration are the
ones that are most attractive and useful to humans—that is, ecosystems containing the resources that fuel our settlement, subsistence, and economic systems. Although most of these places have
been heavily affected by human activities, two island groups provide
worthwhile examples of what control sites can offer: the uninhabited
northwestern Hawaiian Islands and Palmyra Atoll, located about
1,600 km (1,000 miles) southwest of Honolulu.

These island groups have remained largely free from the
impact of commercial fishing. The northwestern Hawaiian Islands
were not occupied when Europeans first arrived in the eighteenth
century, and today they are part of the Pacific Remote Islands
National Monument, which restricts access by tourists and scientists. Similarly, Palmyra Atoll has seen very little human occupation or disturbance and is a designated US national wildlife refuge.
Certainly the seas around these islands have been fished and the environment has been altered by humans and by the climatic changes of
recent decades, but the effects of human activity have been much less
marked than in most places. Along our nature-culture continuum,
these islands are more natural than cultural.

When marine ecologists set out to quantify the differences
between these control islands and other, less protected ones, the
results were shocking. The distribution of large versus small fish

around the main Hawaiian Islands was compared to the fish populations of the northwestern Hawaiian Islands and Palmyra Atoll.
Large, carnivorous fish (the most commercially desirable species) made up only 3 percent of the fish biomass around the main
Hawaiian Islands but 54 percent around the northwestern Hawaiian
Islands. The differences around Palmyra Atoll were even larger. The
number of large fish—tuna and sharks in particular—astonished
scientists and could not have been predicted based on current data
from fish populations around the world. Half a century of fishery
closures and ecological protection and recovery along these islands
resulted in an ecological abundance that most scientists did not
believe possible. We have become so accustomed to an ocean that
is depleted, polluted, and overfished that the ecological richness
of these remote, protected atolls was unimagined. This example
should remind scientists, and all of us, that our natural world has
been so fundamentally degraded and altered by humans that it is
difficult to even imagine what it was once like.

While control sites can show the scale of changes that have
taken place in our world, largely as the result of human actions,
there are precious few such examples. There is no control site for the
deforested jungles of Southeast Asia, for the overhunted and over-
grazed plains of East Africa, for the overfished waters of the North
Atlantic. Our present world holds few clues as to how the world used
to be and how it should look in the future. Where, then, do we look
for information that can help us avoid the shifting baselines syndrome? The answer is to look to the past.

In many scientific fields, the past plays an important role in
advancing the discipline. All knowledge is cumulative, and historical discoveries and revelations help us make new discoveries and
reveal new truths. For example, foundational studies in astronomy
helped us understand Earth, its shape, and its relationship to the

cosmos. Back in the third century BCE, Eratosthenes, the director of the Great Library of Alexandria in Egypt, devised a means through observation to demonstrate that Earth was not flat but round. Eratosthenes read that in the town of Syrene in far southern Egypt at noon on the summer solstice (now observed on June 21), vertical columns cast no shadow. The Sun's rays were perpendicular to the ground. He observed that on the same day in Alexandria, to the north, columns did cast shadows, meaning that Sun's rays hit the columns at an angle in this location. The only possible explanation was that the surface of Earth was curved. Eratosthenes' pioneering astronomical research was the foundation for many other discoveries, including Copernicus's model of the Earth revolving around the Sun; Galileo's confirmation of the spherical shape of the Moon, Venus, and the Sun by observations with a telescope; Kepler's laws of planetary motion; and Newton's laws of inertia and gravity.

The application of historical data in ecology and restoration biology has been much less common than in other fields such as astronomy. While history and the past offer a vast number of control sites against which to measure the current state of the world, environmental data from the historical sciences such as archaeology, geology, and history can be fraught with practical and epistemological challenges and difficult to apply to the present and future. Therefore, most ecologists and biologists have either ignored or overlooked historical data and its relevance for modern resource management. Since historical data were not collected in a systematic fashion consistent with modern scientific practices, these data often are dismissed as speculative or imprecise, and it is impossible to reconstruct historical datasets to conform to modern standards. (Likewise, data collected today likely will not meet the scientific collection standards of the future.) That does not mean, however, that historical datasets cannot help us reach important truths about our modern and future world.

Another challenge in the use of historical information is a lack of communication and interdisciplinary research in scientific fields. This may sound strange: don't scientists always work cooperatively and collaboratively to solve problems and answer questions? Yes, but not always across disciplinary boundaries. Traditionally, scientific disciplines have been separate and specialized. Someone trained as an ecologist or biologist does not necessarily understand how other scientists gather and interpret data: for example, they may not know how archaeologists do their work or why they come to the conclusions they do. Many scientists are now working collaboratively across disciplines, but it takes time, hard work, and creative thinking. And although some scientific fields are beginning to incorporate historical data into current research, this practice remains the exception rather than the rule.

A variety of disciplines gather historical data that can help us interpret and address modern environmental issues. These data, which offer diverse perspectives into past ecological conditions at different time scales—from decades to centuries to millennia—are collected using an array of methods. Examples of short- and medium-term historical datasets include scientific surveys, living memory, and fishing and hunting records; longer-term datasets include written, archaeological, paleontological, and genetic data. Each type of data has strengths and weaknesses, and collecting and integrating different datasets can be challenging. How, for example, should scientists integrate travel logs from fifteenth-century explorers describing the vast abundance of sea turtles in the Caribbean with studies of sea-turtle genetic histories? One dataset is anecdotal, the other quantitative. How reliable or convincing is Ferdinand Columbus's description, during his father's fourth voyage to the New World, that "the sea was thick with turtles so numerous it seemed the ships would run aground on them and were as if bathing in them"? Should such accounts be dismissed as romantic musings, more fiction than science, or can they provide useful information on the biological or physical characteristics of an area? If we take the time and care to collect historical datasets and interpret historical information that can come in many different forms, we can glean important insights about past ecosystems.

HISTORICAL DOCUMENTS

Investigations into the history of ecosystems and efforts to integrate history into our understanding and interpretation of modern environments have been called *historical ecology*. Historical ecologists use a wide array of methods and data to try to understand past and present human-environmental interactions and to reconstruct ecosystems before and after human arrival. Three types of information have become especially valuable. The first is historical documents, including written accounts and pictorial sources. These include reports by naturalists, travelers, and government officials, logbooks from trading and commerce, tax records, newspapers, cookbooks, restaurant menus, maps, illustrations, and paintings. Such documents are often hidden away in libraries, archives, and museum collections, and it requires dedicated sleuthing to uncover the records, compile the data, and apply them. Loren McClenachan, a marine ecologist at Colby College, put together photographs of all the winners from a long-running sport-fishing competition in the Florida Keys, with their trophy catches. By comparing the photographs, McClenachan documented a dramatic fifty-year decline in the average size of the winning fishes, and shifts in the species composition from large groupers to small snappers. Her study is an excellent demonstration of generational amnesia, our shifting views of what constitutes a trophy fish, and the effects of overfishing on large-bodied fish. In a similar fashion, the University of Arizona geoscientist Paul Martin and the archaeologist Christine Szuter combed through the travel journals of Meriwether Lewis and William Clark from their voyage across the American West from 1804 to 1806 for references to large game animals such as bison, elk, and deer. These tell us where these animals once roamed and provide clues as to their former numbers.

ARCHAEOLOGY

Whereas historical records reach back decades or centuries, archaeological records go back millennia. Archaeologists study the plant, animal, and human remains left by ancient peoples in caves, rock shelters, abandoned villages and cities, and trash piles (middens). These data are used to draw inferences about locally available wild and domesticated plants and animals; the spatial and seasonal distribution of plants and animals; the sizes, ages, and relative abundances of animals; the technologies humans used to exploit their environments; and more. Archaeological sites, combined with high-resolution dating techniques such as radiocarbon dating, offer abundant information about past ecological conditions and shifts in ecosystems over long time periods.

Archaeology has provided information valuable for developing fire-management strategies in California. We view wildfires as natural disasters and have instituted fire-suppression policies throughout much of the American West to preserve "natural" ecosystems and protect human settlements. History tells us, however, that before European contact, California was a fire-adapted landscape. Indigenous hunter-gatherer populations intentionally set frequent, low-intensity fires to enhance the productivity of economically important plants and animals, control insects and pests, clear brush, and create pathways. Thus more recent fire-suppression strategies, which often do not include controlled burns, may contravene

An archaeological shell midden site on San Miguel Island, a remote island off the Southern California Coast, which was created by Chumash Native Americans over four thousand years ago. The site is full of valuable cultural and environmental information about life in Southern California millennia ago. (Author photograph.)

millennia of human-environmental interactions. Native flora and fauna might benefit from low-intensity, regular burning. By adopting such policies, we might be able to mitigate the increased intensity and frequency of the wildfires that have devastated swaths of North America in recent years, a direct consequence of anthropogenic climate change.

In another project designed to apply archaeological perspectives to modern management, the archaeologist Iain McKechnie and colleagues synthesized ten thousand years of data on Native American Pacific herring fishing from Washington State to southeast Alaska to assess the health of modern herring populations and the effects of industrialized fishing. They identified an archaeological herring fishery that remained robust and sustainable for millennia with no sign of overexploitation or stress on local populations despite tremendous fishing pressure. Modern herring populations along the Pacific Northwest coast, by contrast, have been fished to the brink of collapse. Archaeological data suggest that an intensive fishery for Pacific herring can be sustainable, but modern populations need to be better managed and informed by Indigenous conservation strategies.

PALEONTOLOGY

Looking even further into the past, we can draw on paleontological data to inform modern environmental management. Paleontologists consult datasets that span thousands to millions of years. Most commonly their data come from rock strata, sediment layers (created by the weathering of parent materials into clays, sands, and other detritus), laminated ice in glaciers and ice sheets, and other geological sources. These data tell us about conditions on Earth not only during human history but also prior to human evolution and the human colonization of pristine environments. Thus they can help determine ecological baselines preceding human influence on an ecosystem and track the evolution of land- and seascapes after human arrival.

Sedimentary records contain traces of flora and fauna such as fossils, shells, plant seeds, pollen, fish scales, charcoal, and phytoliths (fossilized particles of plant tissue). These can help paleontologists reconstruct local, regional, and global environments at different points in time. New analytical techniques examine minerals, trace elements, radioactive isotopes that decay at a constant rate over long periods of time, and ancient DNA. On land, the growth bands (tree rings) of some long-lived trees can tell us about rain-fall, temperature, and other regional environmental conditions; in much the same way, growth bands of corals and other long-lived marine organisms can tell us about past ocean temperatures and salinity levels. Fossils and preserved fauna in sedimentary records provide information on when species first appeared, if or when they went extinct, and how their populations fluctuated over time. Especially valuable are data that show floral and faunal community changes before and after human arrival. Historical ecologists can distinguish changes driven by natural climatic fluctuations from those resulting from anthropogenic influences.

In their study of Chesapeake Bay's eastern oyster fishery over the last eight hundred thousand years, the Smithsonian archaeologist Torben Rick and colleagues examined and measured nearly fifty thousand oyster shells. Their study was designed to investigate the catastrophic collapse of the Chesapeake oyster fishery over the last 150 years as a result of overfishing, introduced diseases, sedimentation, and eutrophication (a high concentration of plant nutrients in water, leading to algal blooms that block sunlight, consume available oxygen, and can cause die-offs of native species). The researchers not only studied the effects of Native American and recent Euro-American oyster harvesting but also contextualized these impacts by studying Pleistocene oyster reefs that existed long before human harvest. They looked for correlations between oyster size and natural climatic changes, such as sea-level and salinity fluctuations. While oysters were, on average, smaller than in the past, the fishery continued to be vibrant and intensive over the thousands of years of Native American harvest. And although the population of Chesapeake oysters is now just 1 percent of what it was just a century ago, over a longer term the fishery demonstrated remarkable resilience to intensive fishing, rising sea levels, and shifting climates. Indigenous foragers exclusively engaged in nearshore oyster harvesting and left deeper water beds to replenish those that were intensively fished, they adhered to strict seasonal rounds, and they focused on immediate consumption by communities in the local area rather than wide-scale trade. These findings offer hope for recovery if we can institute a modern fishery that mimics the relatively sustainable one developed by Native American communities.

With a mix of data, methods, and interdisciplinary perspectives, exploring the history of environments and ecosystems can help us reimagine the environmental crises we face and find ways to address them. By reconstructing the abundances, distributions, densities, and food-web links of species through time and studying climatic shifts, forest cover, ocean conditions, and other factors, we can trace the history of ecosystem changes, as well as the causes and consequences of these changes. This can be challenging work. The good news is that we have lots of historical data; most of it has just been brushed aside in a strict view of what constitutes scientific evidence. All that is needed is a change in scientific attitudes toward historical datasets and a better understanding of what they have to offer.

For example, while most early explorers, traders, ship captains, and merchants, were adept at observing and measuring some aspects of the natural world, such as tides, currents, and weather conditions, their primary goal was not necessarily to record detailed records of the plants and animals they encountered. Yet they left maps, journal entries, and logbooks that document the world from the dawn of globalization. The fourteenth-century Chinese explorer and mariner Zheng He sailed his massive fleet of ships to regions of Southeast Asia, the Indian Ocean, and the east coast of Africa, recording a vast and diverse set of cultures, flora, fauna, land, and seascapes. In the nineteenth century, Robert Fitzroy, the captain of the *HMS Beagle,* on which Charles Darwin sailed as his companion, meticulously documented finch specimens (not true finches but birds belonging to the tanager family) on the Galápagos Islands. The American captain Charles Scammon recorded abundances and distributions of the gray whales he helped drive to the brink of extinction. This information can help us understand the health and distribution of bird and whale populations throughout the Pacific today.

We need to approach this kind of information with caution. Explorers sometimes exaggerated their discoveries to gain funding for future expeditions. Captains had reason to falsify their logbooks to avoid competition and circumvent government restrictions. But modern scientific data are likewise messy, noisy, politicized, and fraught with complications.

A number of scientists are examining these less-traditional sources and considering how they might broaden our understanding of modern ecologies. At times, historical ecology yields shocking and perspective-shifting discoveries. In south Florida, for example, historical data from museum collections and other sources demonstrated that a particular species of balloon vine, once believed to be an invasive weed, is a plant native to Florida and offers critical habitat to the endangered Miami blue butterfly. Efforts to eradicate balloon vine threatened to sever important biotic interactions and push the butterfly even closer to extinction (see chapter 5).

In calling for a better understanding of historical ecology, I am not advocating for turning back the clock. The past is gone; the world has changed. Returning any part of the planet to a previous ecological state is an unrealistic and impossible goal, not least because we can never know exactly what it was like. We can never experience, study, or know the past in the same way as the present.

Although historical ecology helps us construct a picture of the past and answer questions about how we arrived at the present state, we can never re-create that past. This does not mean, however, that what we can learn about the past is irrelevant.

Historical scientists approach their studies with slightly different goals and expectations from other scientists. They are often more willing to sacrifice some degree of precision for general applicability and realism. Most ecologists, biologists, and resource managers, for example, rely on quantitative estimates of population size to track the health of modern species. Fluctuations in population sizes of a few percentage points may be very significant over short time periods. Over longer time scales, however, such changes may be much less significant. Historical data rarely provide quantitative data as specific as data from modern studies, but they allow us to see broader patterns.

All scientists construct some form of narrative about their observations and interpretations and their significance, but historical scientific narratives are distinctive in that they focus on trends over time and describe patterns, causations, and changes as part of interconnected events. They offer descriptions and comparisons rather than quantitative data. This is not to say that historical sources do not and cannot produce reliable quantitative data; sources ranging from tax records and logbooks to shell middens and fossil beds offer many varieties of numerical observations. Yet historical analyses always hinge on incomplete evidence. If we are to employ such records to help tell a story, we need to become comfortable with some level of uncertainty. With the sacrifice of precision comes the opportunity to gain a richer understanding of how the world once was, how it became the world we know today, and where it might be headed.

FORWARD INTO THE PAST

Rather than fill the pages of this book with numbers, charts, figures, and data about the past, I am going to try to tell a good story. To do so, I draw on the work of scientists across many disciplines—biologists, ecologists, historians, archaeologists, geographers, paleontologists, chemists, geologists, and climatologists—to weave together a compelling narrative about environmental change, and about why and how we must look to the past to understand the present and chart our future.

A group of scientists, mostly geologists, recently proposed that the International Geological Congress add a new epoch to the geological time scale, which divides the history of the Earth into distinct, named periods. The Mesozoic Era, for example, extended from about 245 to 66 million years ago. It is divided into three periods: the Triassic, Jurassic, and Cretaceous. The group proposed adding a new epoch called the Anthropocene, with a starting date of 1950. The word *anthropocene* is derived from the Greek roots *anthropos* (man) and *cene* (new)—the new age of humans. The term was popularized by the chemist Paul Crutzen and the biologist Eugene Stoermer over twenty years ago, when they argued that the most recent geologic epoch had ended. What makes this proposed addition so significant is that it identifies humans as the principal agents of Earth system change during this period.

The 1950 starting date was selected because of the initial release of atmospheric radionucleotides, which are highly recognizable in global atmospheric records. These were part of a dramatic increase in human activity affecting the planet, dubbed the Great Acceleration. Human-induced alterations to Earth's ecosystems since that date can be observed in the form of rising greenhouse gas levels, ocean acidification, deforestation, biodiversity loss, and

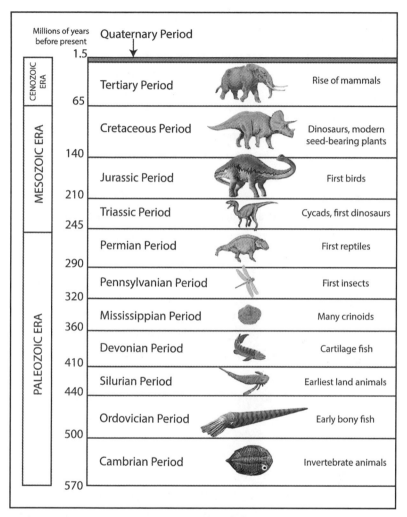

Millions of years before present

	Quaternary Period ↓		
1.5			
Tertiary Period			Rise of mammals
65			
Cretaceous Period			Dinosaurs, modern seed-bearing plants
140			
Jurassic Period			First birds
210			
Triassic Period			Cycads, first dinosaurs
245			

CENOZOIC ERA · MESOZOIC ERA · PALEOZOIC ERA

- **Permian Period** — First reptiles — 290
- **Pennsylvanian Period** — First insects — 320
- **Mississippian Period** — Many crinoids — 360
- **Devonian Period** — Cartilage fish — 410
- **Silurian Period** — Earliest land animals — 440
- **Ordovician Period** — Early bony fish — 500
- **Cambrian Period** — Invertebrate animals — 570

Geological time periods from 570 million years ago to the present (moving bottom to top). The three geological eras are labeled at the left and the twelve periods at the right. The pictures show representative animals of each period (save the Quaternary), along with biological hallmarks of each geological division. Geological periods are labeled with their boundary dates in millions of years ago. The Quaternary period includes the last 1.5 million years of the Cenozoic era. The last 11,700 years of the Quaternary are known as the Holocene, the current geologic epoch. (Images via Wikimedia Commons.)

landscape alterations (such as rerouting waterways, freshwater usage, and fertilizer application). There is not much to quibble with here. It's indisputable that humans have had an increasingly devastating effect on Earth systems in the past seventy-five years. New evidence seems to surface every week—an oil spill in California, record-breaking heatwaves in Europe and Asia, deforestation in the Amazon, megafires in the American West, and Pacific islands being submerged by sea-level rise. The Anthropocene concept is a real and important idea. Humans have created an imperiled Earth, a planet at grave risk of ecological collapse. More than ten thousand scientists signed an initiative in 2019 declaring a worldwide human-induced climate emergency and calling for urgent action to tackle it.

If we believe that human activity began to affect the planet adversely only during the past one hundred years or so, we might conclude that earlier history has no lessons to offer for constructing policies to mitigate our climate and environmental crises. Nothing could be farther from the truth. History matters in the Anthropocene. In chapter 2, I set the stage by tracing how, some fifty thousand years ago, humans began spreading across the globe, effectively adapting to and shaping the ecosystems and landscapes they encountered in order to meet their needs. Chapters 3–6 offer specific examples of how history matters for understanding and mitigating our environmental crisis. These chapters address the four greatest, and interconnected, challenges facing our world: anthropogenic climate change, deforestation, declining biodiversity, and overfishing. There are certainly other environmental problems, such as pollution, human overpopulation, and access to freshwater. But if we could control these big four, we would be well on our way to building a more sustainable and resilient world. If we can curb deforestation, we can increase biodiversity and reduce greenhouse gases. If we can better manage human impacts on the oceans, kelp forests and sea grasses will prosper. These organisms sequester greenhouse gases and help marine ecosystems thrive.

Each of these chapters starts with a current event and travels back in time, demonstrating how archaeology and the historical sciences can be applied to modern sustainability efforts. In the epilogue, I return to the Anthropocene, exploring what it means to live in the age of humans. I take a close look at the modern environmental crisis, the challenges facing the global community, and the ways archaeology and history can help address some of our most pressing problems.

Building a healthier and more sustainable planet not only benefits the plants, animals, and natural places of our world but can also do much to benefit humankind. Healthy human communities depend on well-functioning ecosystems for clean air, fresh water, medicines, and food security. A healthy world helps humans live longer and happier lives; it enhances physical, mental, and social well-being. Addressing our environmental challenges and building sustainable systems is an integral part of creating a more equitable and inclusive world. In innumerable ways, we can all benefit from confronting the environmental challenges of imperiled Earth.

○

Chapter 2

OUT OF AFRICA
AND AROUND THE GLOBE

T ime can be difficult to put into perspective. Episodes in our lives from just a year or two ago may seem distant memories. Conversely, our daily routines can make one year blend into the next, making five years ago seem like yesterday. Events preceding our lived experience can be difficult for us to fathom. Coming across a World War II documentary or finding early-twentieth-century photographs in family albums leaves me awed at how much the world has changed. I've never lived during a time where the planet's most powerful nations were in conflict. It's difficult for me to imagine what life would have been like before the invention of telephones and automobiles.

If it's hard to imagine life just a century ago, it's far more difficult to think about the history of our planet. Consider the history of dinosaurs on Earth. The Jurassic Park movies give the impression that the age of giant reptiles was a relatively short and contained interval in Earth history. In reality, dinosaurs dominated the land and sea for 180 million years, spanning multiple geologic epochs. They survived and thrived for thirty-six thousand times longer than humans have existed.

In daily life we tend to think in short time frames—days to years. Geoscientists think on a scale of hundreds of millions of years, considering how the processes of erosion, plate tectonics, and other slow-acting geological forces have brought about major changes to our planet, along with sudden, infrequent events such as floods, volcanic eruptions, and asteroid impacts.

Earth's history extends back 4.5 billion years; it came into existence approximately 10 billion years after the universe formed. Life on our planet first emerged some 3.8 billion years ago as tiny microbes, identified by fossils preserved in rocks. The continents drifted across the globe, the chemical composition of our atmosphere shifted, climatic patterns fluctuated, ocean currents changed, and plants, animals, and ecosystems evolved and went extinct.

In the geological history of planet Earth, humans are a footnote, a blink of an eye. The first ancestors of modern humans appeared about seven million years ago. These human ancestors, called *hominins*, were distinguished from other primates by their upright posture, bipedal locomotion, larger brains, and innovative behavioral characteristics. Our ancestors learned to create tools, control fire, build shelters, cooperate, and adapt to shifting environmental conditions. In the late nineteenth century, Charles Darwin recognized that Africa was probably the epicenter of human evolution, a point he made in his 1871 book *The Descent of Man.* Early hominin fossils unearthed in Africa tell the story of humans' anatomical and technological evolution, popularized by the famous paleoanthropologists Louis and Mary Leakey, among others. The Leakeys discovered and named several early hominin species in Africa, some dating back millions of years, and helped reconstruct the evolutionary origins of humans.

By no means do we know everything about the evolutionary history of humans. In my introductory archaeology and prehistory class at San Diego State University, I spend the first seven weeks reviewing highlights in human physical and cultural evolution. I do my best to keep up with the ever-changing landscape in paleo-anthropology, but it seems that every few months there is a new hominin fossil, an innovative ancient DNA study, or some other discovery that alters the story. Most scientists recognize between fifteen and twenty different species of early human ancestors, and debates rage about how these species are related, how to identify and classify them, and what factors influenced their evolution and extinction.

Most paleoanthropological (i.e., fossil) and genetic evidence suggests that *Homo sapiens sapiens,* or anatomically modern humans (AMHs), the species to which we belong, first evolved in Africa approximately two hundred thousand years ago, although some evidence suggests that this chronology might extend back another one hundred thousand years. This new species differed from archaic (premodern) *Homo sapiens* with relatively minor changes in the cranial anatomy, a slightly larger brain, smaller teeth, and the more gracile bones. These differences were probably relatively subtle. If you sat next to an archaic human ancestor dressed in a suit and tie on a bus, you probably would not notice anything out of the ordinary.

Like earlier hominin (direct human ancestors) species, AMH populations began to migrate out of their African homeland, inter-acting and mixing with other hominin species in Europe, the Middle East, and Asia. They successfully adapted to new environments and ecosystems and developed new tool technologies to hunt and gather Pleistocene flora and fauna.

Widespread evidence shows that by around fifty thousand years ago, AMHs were creating jewelry and personal ornamenta-tion, figurative art such as cave paintings and elaborately carved human and animal figurines, and early forms of religion or cosmological thinking. AMHs developed new hunting technologies, formed larger, cooperative hunter-gatherer settlements, and likely developed sophisticated languages. This transition to behavioral modernity marks what archaeologists call the Upper Paleolithic Age. AMHs not only looked like us but acted like us. Increased cooperation, cognitive flexibility, and technological innovation allowed our species to populate the world and form the complex societies we know today.

COASTING INTO NEW WORLDS

The development of new tools, technologies, and adaptive behaviors helped propel AMHs to new geographic frontiers, starting around 185,000 years ago. They adapted to the new environments of Europe, the Middle East, and Asia. One migration route, known as the Southern Dispersal Route, followed the southern coast of Asia, from the Arabian Peninsula to India and on to Southeast Asia and Oceania. Although several other hominin species already occupied locations outside Africa, it was Upper Paleolithic AMHs who flourished and eventually replaced all other hominins.

Along the Southern Dispersal Route, AMHs likely followed sheltered coastlines. They would have experimented with watercraft and explored various coastal environments, including mangrove forests, with their rich mix of terrestrial and marine flora and fauna. Unfortunately, there has been very little systematic archaeological research in these regions. As a further complication, since organic materials such as skins, wood, and plant fibers are not well preserved in the archaeological record, we have no idea how and when the earliest watercraft were constructed or how their manufacture evolved. Moreover, sea-level rise has submerged many former coastlines and may have drowned much of the archaeological record of early coastal migrations.

DNA studies and archaeological investigations seem to support the idea that after leaving Africa, AMHs followed South Asian coastlines to the shores of Australia between about eighty thousand and fifty thousand years ago. The physiography of the coastlines would have been quite different in this period. The lower sea levels of the Ice Age exposed now-drowned coastal plains. Many Southeast Asian islands were connected into a single landmass (called Sundaland), water gaps between islands were narrower, and a supercontinent known as Sahul encompassed Australia, Papua New Guinea, and Tasmania. Running between Sundaland and Sahul were deep ocean trenches with strong currents. At least one of these crossings was nearly one hundred kilometers (about sixty-two miles) wide and would have required multiple days of travel. Sometime between sixty thousand and fifty thousand years ago, the ancestors of Australian Aboriginal peoples arrived on the shores of Sahul and quickly dispersed along the continent's coastlines and waterways. This remarkable accomplishment, the first fully maritime voyage in human history, opened a new continent for human settlement.

The peopling of the Americas likely played out in similar ways to that of Australia, although much later. Until relatively recently, most archaeologists believed that the first Americans walked across the Bering Land Bridge (the now-submerged land that periodically connected Siberia and Alaska). As the two massive ice sheets that covered much the North American continent slowly melted, the migrants passed through an ice-free corridor and rapidly spread across the Americas beginning about 13,500 years ago. Known as the Clovis peoples, they were believed to be specialized hunters of New World megafauna (animals that weigh more than

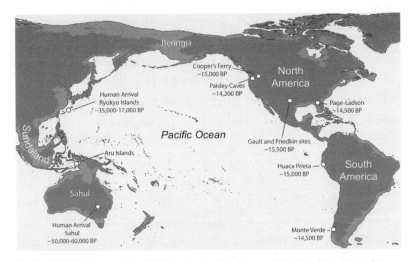

The Pacific Rim at the Last Glacial Maximum, approximately twenty thousand years ago, when sea levels were lower across the globe. Labeled are the dates of human colonization of Sahul and the Ryukyu Islands and some of the pre-Clovis sites in North and South America. Given that massive ice sheets blocked the overland passage to North America until about fourteen thousand years ago, the earliest humans likely came to the Americas in boats along the Pacific coast. (Base map via Wikimedia Commons.)

forty-four kilograms, or ninety-seven pounds), such as mammoths and mastodons. This story about the peopling of the New World is the one with which most people are familiar. It is certainly the story I learned in high school and college, and at the time, the vast majority of archaeological evidence seemed to support it. But it's simply not true.

For decades, the Clovis-first model of New World settlement dominated American archaeology. This all changed in the 1990s, when a 14,500-year-old site was discovered in southern Chile that predated Clovis culture and the opening of the ice-free corridor by 1,000 years. This uniquely well-preserved archaeological village site, Monte Verde, has produced a rich array of tools, subsistence remains, and the remnants of tent stakes and hearths all very different from typical Clovis sites. At the time of its discovery, most archaeologists believed that people did not turn to coastal fishing and gathering until Ice Age megafauna went extinct about ten thousand years ago, forcing them to modify their hunting lifestyles. Monte Verde has produced evidence of coastal foraging, including four different types of edible seaweed, and a diverse subsistence economy. Genetic studies and the unearthing of several other widely accepted pre-Clovis sites, such as Paisley Caves in Oregon and the Gault and Friedkin sites in Texas, have shattered the Clovis-first paradigm. Questions remain about how and when this migration took place, but the first Americans likely arrived in the New World in boats, following Pacific coastlines like the first migrants to Sahul.

With their arrival in the New World, AMHs had reached every continent on the planet, save Antarctica. More than ten thousand years would pass before humans made their final push of discovery to the islands of the Pacific Ocean. The initial exploration and settlement of the Pacific Islands is one of the most remarkable accomplishments in human history. The Pacific covers more than one-third of the Earth's surface, stretching 15,500 kilometers (about 9,600 miles) from north to south and 20,000 kilometers (about 12,400 miles) from east to west. Only 0.7 percent of this vast oceanscape consists of islands, which are on average sixty square kilometers (twenty-three square miles) in size. Island groups are often separated by vast stretches of open ocean. Yet people managed to discover, explore, and colonize these islands.

The first wave of Pacific colonization began with the larger islands of Melanesia, such as the Bismarck Archipelago, by about thirty-five thousand years ago. This was followed by an eastward movement into Buka in the Solomon Island chain around twenty-eight thousand years ago. These migrants were hunter-gatherers with sufficient navigational skills and maritime technologies to cross

The discovery and peopling of the Pacific Islands. The two large arrows at the lower left show early waves of migration by hunter-gatherer groups into Australia and Papua New Guinea starting sixty thousand to fifty thousand years ago. Arrows from Taiwan to the Philippines and out into the Pacific depict migrations by Polynesian peoples, extending into the Polynesian triangle. (Map via Wikimedia Commons.)

the 180 kilometers (112 miles) of open water between the Bismarck and the Solomon islands. At this point migration stopped, however, and populations settled along the coastlines and forested interiors of the islands. It was not until the second migration push, starting about 3,500 years ago, that the ancestors of Polynesian peoples discovered and colonized the many islands of the central and southern

38

Pacific. Unlike the first wave, these migrants were agriculturists who brought with them domesticated plants and animals, including dogs, pigs, chickens, and root crops, and new maritime technologies such as seafaring canoes and fishing tackle, along with distinctive pottery. These migrants systematically explored the Pacific, settled the habitable islands they encountered, and reached Hawaiʻi, Rapa Nui Easter Island, Aotearoa New Zealand, and nearly every island in between by about one thousand years ago.

With this second migration push, humans began to reshape local and regional environments and ecosystems. These changes had both positive and negative effects on the resilience and structure of ecosystems. Studying these outcomes can help us understand more recent anthropogenic environmental changes. One of the earliest changes driven by humans was the extinction of megafauna around the world between about fifty thousand and ten thousand years ago. Human effects on the environment changed and accelerated in concert with changes to subsistence, settlement, and sociopolitical systems.

HUMANS AND MEGAFAUNAL EXTINCTIONS

As humans radiated from Africa, their presence triggered ecological shifts and changes. The first AMHs to arrive in these new landscapes were small groups of highly mobile hunter-gatherers, who probably left a light ecological footprint; they certainly left only an ephemeral archaeological record. It may be, however, that they contributed to the extinction of some animals.

During the first arrival of humans to Australia and the Americas, the world was full of creatures that now sound fantastical. Eurasia was home to giant creatures such as European hippopotamuses, aurochs (a kind of wild ox), cave hyenas, lions, bears, and woolly rhinoceros. North America had lions, dire wolves, and giant sloths; South America boasted two-ton bears and massive camel- and rhinoceros-like creatures. Giant wombats three meters (ten feet) tall and flightless birds the height of professional basketball players inhabited Australia. Woolly mammoths and mastodons roamed all these continents save Australia. Outside Africa, fewer than half of these species of megafauna remain. As human settlement spread, megafauna around the globe began a catastrophic decline, with about 90 of 150 genera driven to extinction by ten thousand years ago.

Explaining why these large animals went extinct remains one of the most controversial topics in archaeology, ecology, and evolu- tionary biology. The two primary explanations are dramatic climate change and human hunting. Megafaunal extinctions between about fifty thousand and ten thousand years ago may constitute the earli- est human-induced biotic crisis in Earth history.

New data, new studies, and new interpretations continue to inform this debate. We know that in Australia, some twenty-one genera—representing about 83 percent of large marsupials, birds, and reptiles—went extinct approximately 46,000 years ago. These included giant kangaroos, wombats, and snakes. In Northern Eurasia, Siberia, and Alaska, nine genera, representing about 35 percent of megafauna species, went extinct in two pulses. Warm-weather-adapted mega- fauna, which include straight-tusked elephants, hippos, certain wild horses, and short-faced bears, went extinct between 48,000 and 23,000 years ago, and cold-adapted megafauna, such as mammoths, went extinct between 14,000 and 11,500 years ago. In continental North America, approximately thirty-four genera (72 percent) of large mammals went extinct between about 13,000 and 10,500 years ago, including mammoths, mastodons, and giant ground sloths; some species of horses, tapirs, and camels; short-faced bears, and saber-tooth cats. Large mammals were most severely affected, but some small mammals, including a variety of skunks and rabbits, also went extinct. South America lost an even larger variety of megafauna, with fifty genera (83 percent) becoming extinct at about the same time as those in continental North America. Overall, across Eurasia, Australia, and the Americas, large-bodied animals with slow reproductive rates were especially prone to extinction.

Many scholars believe that these extinctions are best explained by the rapid climate changes and consequent vegetation shifts, habitat loss, and resource scarcity at the end of the last Ice Age event. In Australia, extinctions of megamarsupials and other large-bodied animals are attributed to increasingly arid conditions. In the Americas and Eurasia, climate change proponents point to severe and rapid warming trends. Grassland and other habitats on which grazing megafauna relied were reduced and so transformed that they could no longer support these animals, and natural migration barriers prevented them from moving in search of food. Supporting this scenario are studies from North America suggesting that the largest herbivore predators (dire wolves, saber-tooth cats, and short-faced bears) went extinct, while species with more diverse diets (gray wolves, puma and bobcats, and brown and black bears) survived.

The competing idea that human hunters were the primary cause of megafaunal extinctions was popularized not by an archaeologist, but by the University of Arizona geoscientist and paleoclimatologist Paul Martin. Using computer simulations and hunting models, Martin argued that the peopling of the Americas at the tail end of the last Ice Age resulted in megafaunal extinctions within a millennium as humans spread from north to south hunting mammoths, mastodons, and other large animals for their meat, bones, and hides and disrupting their reproduction patterns. In similar fashion, he asserted, the arrival of humans in Australia instigated a

wave of extinctions resulting from the human hunting of megamarsupials and other large animals. Martin suggested that these animals were quickly overhunted because they lacked the behavioral
and evolutionary adaptations necessary to avoid novel, intelligent,
and technologically sophisticated human predators. Extinctions
were less numerous in Africa and to a lesser degree Eurasia because
megafauna and human hunting had coevolved, and animals in these
regions had developed instincts and strategies for avoiding humans.

The primary cause of megafaunal extinctions continues to
be hotly debated in the scientific community because both the
climate-change and human-overhunting hypotheses, though they
offer powerful arguments, have shortcomings. At the end of the last
Ice Age, a wide variety of megafauna were certainly lost to extinction
amid rapid changes in climate and environment. However, there
have been as many as twenty climatic cycles of similar proportions
in the past five million years of Earth's history. None of these are
associated with extinctions at the scale and magnitude of those of
the last glacial period. What makes the last ice age different from all
the previous ones is the presence of fully modern humans on every
continent except Antarctica. Supporters of the climate-change theory and opponents of the overhunting theory argue, however, that
very few archaeological sites in the Americas and Australia show
evidence of megafaunal hunting. In all of continental North America, fewer than two dozen sites offer clear evidence of human hunting of megafauna such as mammoths and mastodons. In addition,
as archaeologists have refined chronologies for the initial peopling
of Australia and the Americas, no clear correlation has emerged
between megafaunal extinctions and the arrival of humans.

To my mind, the most likely explanation is that megafaunal
extinctions were driven by a mix of human, climatic, and ecological factors. In the Americas, it is possible that megafauna herbi-

vores such as mammoths and mastodons were already stressed at the end of the Last Glacial Maximum by carnivore predation and climate change. The arrival of human hunters was enough to tip the scale toward extinction, creating a cascading effect of ecological change that led to the extinction of other large and small animals. In Australia, extinctions were driven by a warming climate and Aboriginal fire-stick farming. Also called *cultural burning* or *cool burning*, this is a form of controlled burning used by Aboriginal Australians for millennia to clear vegetation, promote the growth of desirable plants, and attract game. Extinctions may have resulted from a complex feedback loop whereby the loss of large herbivores increased the amount of combustible vegetation, and this shift, combined with increasing aridity, generated more intense fires.

The arrival of fully anatomically modern humans into Australia, Eurasia, and the Americas, whatever their role, was followed shortly by a wave of megafaunal extinctions. In Eurasia, an early extinction pulse coincides with the arrival of modern humans. A second pulse may have been linked to human population increases, increased hunting pressure, and the invention of new tool technologies, but it also coincides with climatic warming and vegetation changes.

The megafaunal extinctions that began some fifty thousand years ago are at least partly tied to human arrival. Since then, plant and animal extinctions and the role of humans in this process have increased dramatically, especially in the last several hundred years. Today, the rate of human-induced species extinctions is thousands of times greater than background rates (i.e., extinctions due to climate change and other natural factors). Modern extinctions and the loss of biodiversity around the world are the consequences of a long and continuing process facilitated by humans. Understanding this history is vital for effective and targeted conservation efforts (see chapter 5).

FROM HUNTER-GATHERERS TO AGRICULTURALISTS

Not long after the last of the Ice Age megafauna disappeared from the Americas, human communities began transitioning from hunting and gathering to agricultural societies. Around ten thousand years ago, domestication of wild plants and animals began in southwest Asia, Southeast Asia, sub-Saharan Africa, Papua New Guinea, and parts of the Americas. This was a watershed moment in the relationship between humans and the environment. It transformed nearly every aspect of human life and was the first step in the growth of large urban centers and the rise of the sociopolitical hierarchies.

The ability to grow, harvest, and store food allowed for technological innovation, the expansion of geographic and scientific knowledge, greater communication and interaction, the conquest of many diseases, longer human life expectancies, and wide access to crops, foods, and other essential goods. But these changes also produced ugly outcomes for human societies, including the spread of infectious diseases, war, pestilence, genocide, exploitation, and the loss of linguistic and cultural diversity. In addition, human food production has taken a large toll on local, regional, and global environments.

As human subsistence economies moved away from generalized hunting and foraging to specialized and intensive agricultural production, ecosystems were transformed. Even in its earliest stages, the agricultural transition resulted in significant environmental changes. Natural vegetation was cleared to create agricultural fields and pastures. Streams and rivers were dammed and diverted to irrigate cultivated land. Human populations increased, fully sedentary communities formed, the translocation of plants and animals accelerated, new diseases appeared and spread, and habitat alterations intensified. Human activity outstripped natural climate change as the driving force behind plant and animal extinctions and local ecosystem change.

As urban centers grew, habitat destruction, land clearance, and other human-induced effects expanded from local to regional and continental scales. Rodents, weeds, dogs, and livestock spread around the globe, competing with and preying on native species. As land was cleared, topsoil was exposed to erosion, affecting landscapes and plant growth.

One particularly illustrative example comes from 'Ain Ghazal, an agricultural village in southern Jordan dating back ten thousand years. 'Ain Ghazal grew from a small hamlet into a settlement spanning more than twelve hectares (about thirty acres) in about five hundred years, but it ceased to be permanently occupied about seven thousand years ago. Many archaeologists and historians have cited decreased annual rainfall and climate change as the primary reasons for the depopulation of 'Ain Ghazal and other early agricultural villages in the region, but decisions by the early occupants may have played equally important roles in the decline of the village.

The occupants of 'Ain Ghazal built substantial homes with stone walls and floors, finished with a layer of lime plaster. Creating lime plaster requires high heat, consuming some four tons of wood fuel per ton of plaster produced. Some houses at 'Ain Ghazal had dozens of plaster layers, suggesting that occupants plastered their homes every few years. Gathering the wood for this activity led to deforestation for a radius of three or more kilometers around the village. In adjacent regions, clay was mixed with water to form mud plaster, or gypsum was heated to low temperatures to form gypsum plaster. Both techniques were fairly sustainable compared to using lime plaster.

'Ain Ghazal also differed from other regions by maintaining flocks of goats rather than sheep. Goats were better adapted to the area's steep topography and more arid climate. Because goats consume coarse vegetation, including seedlings and young saplings, they impede the regeneration of forest. A scarcity of vegetation exposes soils to the elements and increases erosion, decreasing the land's capacity to support plant life. Sheep, on the other hand, are less destructive to forests because of their preference for faster-growing grasses and weeds.

The landscape of the Levant (which includes modern Syria, Lebanon, Jordan, Israel, the Palestinian territories, and much of Turkey) is characterized by deeply cut stream courses. Freshwater is available only near springs and along streambeds, which are dry in some seasons. Agricultural villages were sited near freshwater, on flat terraces or gently sloping surfaces, and agricultural fields fanned out from the village centers. Goats provided useful manure for fertilizing fields, but as forests were felled to provide fuel for making plaster and for the domestic hearth, and goats grazed outside garden plots during the growing season, adjacent areas were increasingly devegetated and unable to regenerate. Fall and winter rains washed away topsoil along steep slopes, permanently altering local vegetation communities.

All these decisions created a devastating feedback cycle. As villages grew, agricultural fields expanded, forests were felled, and soils eroded. Villagers responded by collecting wood from more distant forests and grazing goats farther and farther afield. Native plants and animals became increasingly scarce and increased people's reliance on domesticated flora and fauna, which in turn increased the size and number of agricultural fields, as well as the number of goats. The degradation of the local environment made it increasingly inhospitable. About eight thousand years ago, local climatic change brought about a series of unusually dry years that was the coup de grâce for a number of local villages, although 'Ain

Ghazal persisted for about a century longer because of its proximity to diverse local resources. Gradually people abandoned these villages, fragmented into smaller groups, and adapted a more transient lifestyle.

When we consider what we can learn from 'Ain Ghazal and other early agricultural villages in the Levant, it is useful to try to look at events from the perspective of an 'Ain Ghazal farmer. Gradual deforestation and soil erosion resulting from daily activities—farming, making lime plaster, grazing goats—would be difficult to detect and put in perspective over the span of a single lifetime. Deforestation and erosion happened over centuries. Villagers probably sought ways to lessen the damage and make their agricultural systems work, even as their activities made them more and more vulnerable to environmental collapse. It took years of less-than-average rainfall to force people to abandon their homes. Archaeological investigations demonstrate that people were stressed and actively trying to avoid disaster before this. Houses became smaller, and lime plaster was replaced with crushed lime plaster, which did not require fuel. Ultimately, however, 'Ain Ghazal failed.

The biggest problem for 'Ain Ghazal was tradition. It is difficult to abandon the knowledge and ways of doing things that are passed down to you from your parents, grandparents, and friends. Change is difficult. For the villagers of 'Ain Ghazal to stop grazing goats, to stop constructing houses the way they always had, to stop living the life they had always known, would have posed an enormous and unwelcome challenge. We need only look at ourselves. We know that the unchecked burning of fossil fuels is having devastating and cumulative impacts on our planet. We know that the polar ice caps are gradually melting and raising sea levels. We know that runoff from chemically fertilized agricultural fields is damaging critical freshwater sources. We know the American West is facing unprecedented drought. But in the short term, we can live with these changes. So we cling to our traditions, even when they are increasingly unsustainable and damaging.

DISCOVERING ISLANDS

A number of archaeologists, including me, focus on the history of islands. For those interested in untangling the long histories of human-environmental interactions, islands offer unique advantages. Because of their small size and relative isolation, islands make it possible to observe the impacts of climate fluctuations, human arrival, and human activities more clearly. For example, it is often easier for archaeologists to determine when humans introduced nonnative flora and fauna, hunted or otherwise drove animals to extinction, or modified land- and seascapes. The lessons learned can then be applied to larger, more complicated continental systems.

Each island or island group presents a case study in the history of human-environmental interactions. During the early stages of human settlement, burning, landscape clearance, and the introduction of plants and animals were the primary drivers of ecological change. As populations increased and subsistence and other activities intensified, extinctions, soil erosion, and land and seascape modifications often intensified as well. The trajectory and scale of human impacts depend on complex, intersecting variables, from island physiography to human subsistence strategies, population densities, technology, sociopolitical organization, and human decision-making. In some cases, the impacts were detrimental to island biodiversity and soils, making human settlement unsustainable. In other cases, people arriving on new islands developed successful strategies to manage their populations and create sustainable subsistence systems.

All over the Pacific Islands, land-dwelling birds were heavily affected as maritime agriculturalists arrived beginning some 3,500 years ago. Thirteen of seventeen species went extinct on Mangaia in the Cook Islands shortly after human arrival; five of nine went extinct on Henderson Island in the eastern South Pacific; seven of ten on Tahuata in the Marquesas; ten of fifteen on Huahine in the Society Islands; and all six species on Rapa Nui Easter Island. In the Hawaiian Islands, more than 50 percent of the native birds went extinct after Polynesian arrival but before first European contact (with the arrival of Captain Cook). These extinctions resulted from human hunting, burning, deforestation, and other habitat alterations, along with the introduction of domesticated animals (such as pigs, dogs, and chickens) and accidental stowaways (rats). In contrast, on islands without significant prehistoric occupation, there is little evidence for bird extinctions prior to European arrival.

Many of the most devastating effects of the Polynesian spread across the Pacific probably originated from the unintentional introduction of rats from mainland southeast Asia. Rats likely climbed aboard the large sailing canoes to feed on root crops and other food items in storage, which provided sustenance for weeks to months out at sea. When the canoes made landfall, the rats disembarked along with the human travelers. It would have taken only one pregnant female rat or a small number of individuals to quickly populate an island. The rats easily adapted to ecosystems that had few or no predators to keep their numbers in check. They have been implicated in the extinction of various snails, frogs, and lizards in Aotearoa New Zealand, giant iguanas and bats in Tonga, and a variety of birds across the Pacific. The story of deforestation and environmental deterioration on Rapa Nui Easter Island, attributed by some to the building of the famous Moai statues, has been used as a cautionary tale about the dangers of overexploitation. But the true driver may have been rat rather than human activities. In the early 2000s, the geographer John Flenley identified rat gnaw marks on the seeds of the now-extinct Rapa Nui palm, suggesting that these rodents played a significant role in the extinction of this species and the decrease in island biodiversity, as well as the subsequent lack of construction material for oceangoing canoes and other purposes.

Islands serve as important models for examining the long and complex interplay between humans and their environments—not only the negative impacts of human arrival and subsequent activities, but also the ways people may have enhanced or managed island resources. Island explorers introduced plants and animals, harvested native species, and modified their land- and seascapes in assorted ways. Studying the trajectories of human-induced island change can provide invaluable lessons about environmental problems and the historical efforts to find solutions.

THE PROBLEMS OF COMPLEX SOCIETIES

Over the past five thousand years, one of the prevailing trends in human history has been the rise of densely populated cities. This trend was facilitated by the domestication of plants and animals and subsequent innovations in agrarian food systems that made it possible to produce, store, and transport larger quantities of food. Like the transition from hunting and gathering to agriculture, the growth of cities occurred at different times in different parts of the world. In most regions, agricultural villages were the centers of food production, but cities became the distribution sites for food surpluses; the nexuses of trade and exchange; the incubators of specialized occupations, technology, and innovation; and the seats of government, elite, and religious control. Even so, it was not until the twentieth century that cities became homes to the majority of a region's population.

The growth of urban centers required more intensive exploitation of local and regional resources, lengthening what the sociologist Sing Chew calls the city's "ecological shadow." City populations require large amounts of natural resources, foods, and raw materials, which must sometimes be transported over long distances. They need materials for the construction of homes, government and religious buildings, and infrastructure. They require roads, canals, and ports to facilitate the flow of goods and people to and from cities. As cities grow and multiply, their ecological shadows extend farther and farther, into local, regional, and eventually global landscapes, seascapes, and economies. These features result in increasing ecological stress.

A number of well-known archaeological examples demonstrate the stress and transformative effects of large urban centers on local environments. In Cambodia, for example, the city of Angkor, approximately one thousand years old, is home to elaboratively carved stone temple complexes sprawled over an area of more than 1,000 square kilometers (390 square miles)—nearly four times the size of modern New York City. Angkor was the capital of the Khmer Empire, which dominated much of Southeast Asia from the ninth to the fifteenth century CE, and the largest preindustrial city in the world, with an estimated population of nearly one million people. Using airborne laser scanning technology called LiDAR, which can penetrate dense jungle vegetation and map the underlying terrain, archaeologists have meticulously reconstructed Angkor's temple complexes, roads, canals, moats, and ponds. These efforts show that rather than occupying a densely populated city core, residents were distributed throughout the sprawling city complex in small villages. Dwellings were connected by roads and irrigation networks that facilitated the movement of products and people across the city and provided water for extensive plots for growing rice and crops. In addition, Angkor boasted four massive, manmade water reservoirs, called *barays*, which had storage capacities of between 7.5 million and 48 million cubic meters (about 2 billion to 12.6 billion gallons) of freshwater. That is enough to fill about 40 million standard American bathtubs.

Angkor's hydrological network was a wonder, an ingenious adaptation to challenging environmental conditions. Angkor lies in the lower Mekong River Basin. It is subject to annual monsoon cycles, which create a rainy summer monsoon season followed by a dry season often characterized by severe drought and ecological stress. Annual rainfall in the area is about 1,400 millimeters (55 inches) per year, with consistently warm temperatures. The notoriously gloomy and wet Pacific Northwest cities of Portland, Oregon, and Seattle, Washington, by comparison, average slightly under 1,000 millimeters (36–38 inches) of annual rainfall. During the monsoon season, violent storms can release massive amounts of water over short periods. Large areas of the Angkor plain can become inundated. The Khmer people understood the need to respond to these climatic patterns. The city's extensive waterworks were designed to manage seasonal changes in water availability. Residents diverted local rivers to create moats and fill *barays*, allowing them to grow rice and other staple crops not only during the rainy season but year-round. Complex canal networks helped control flooding and capture and distribute surface water. Combined with the spectacularly rich fishery of the local Tonlé Sap Lake, the control of water for irrigation and transportation enabled the Khmer Empire to dominate Southeast Asia for six centuries.

After surveying these aquatic features alongside house mounds and clusters of shrines across the ancient city, the French archaeologist and architect Christophe Pottier commented, "The

The main palace of Angkor Wat, built in the early twelfth century as the seat of political and religious power of the Khmer Empire. The water in the foreground is part of a massive moat, one of the many hydrological feats of the Khmer. (Image via Wikimedia Commons.)

people of Angkor changed everything about the landscape. It's very difficult to distinguish what is natural and what is not." The cityscape, the waterways, the agricultural fields, and the temple complexes constituted a vast cultural landscape—an environment and ecosystem intensively engineered by humans over centuries to meet their needs and adapt to climatic challenges.

This human footprint and environmental engineering came at a price, however. The Khmer Empire fragmented, and Angkor was largely abandoned in the early fifteenth century. The precise causes are debated. Conflict with groups along the empire's borders and political and religious upheaval likely played a role in the collapse, but a combination of climate change and the mechanical break-down of the city's waterworks were also significant factors. Climate data suggest that monsoons weakened during the early centuries of the Little Ice Age, between about 1300 and 1600 CE, bringing less rainfall. When residents could no longer rely on summer deluges to refill canals, ponds, and *barays*, rice yields decreased. Booming populations and increased forest clearance for agricultural fields likely accelerated the erosion of topsoils, which washed into the canals. Archaeologists have identified evidence of failed spillways and accumulating silt in waterways and *barays*, impeding irrigation and contributing to the city's collapse.

Angkor was not the only ancient city-state that unraveled as a result of environmental change and ecological stress. Devastating droughts and human-induced landscape stress are likely responsible for triggering the Maya people's abandonment of much of their lowland homeland on the Yucatán Peninsula in Mexico between 1,100 and 1,200 years ago. There are certainly lessons we can learn from these collapses. News stories about failing American infrastructure—such as collapsing roads, rusting bridges, and contaminated water supplies—offer uncomfortable analogies to the decline of these past cultures. History tells us that the rise of urbanized human societies placed tremendous stress on local, regional, and global environments. Environmental changes effected by humans over millennia helped create the world we know today.

HUMAN LEGACIES IN THE NATURAL WORLD

For at least fifty thousand years, ecosystems and plant and animal communities have been shaped and reshaped by human hunting, gathering, landscape modification, and related activities. This story may be profoundly different from the ones you learned when you were young. I distinctly remember the television commercials of my youth featuring Iron Eyes Cody, an Italian American actor in stereotypical Native American garb, paddling a birch-bark canoe on perfectly calm waters in a tree-lined wilderness. Dramatic drumbeats and pounding rhythms sound in the background. At first the setting seems tranquil and pristine, but as he gently paddles forward, the waters, and shoreline become increasingly polluted, and a massive ship and dock loom in the background. Cody pulls his boat onto the shore and walks to the edge of a bustling freeway. A white paper bag hurled from a passing car bursts into a shrapnel of half-eaten burgers, French fries, and fast-food wrappers that cover Cody's beaded moccasins. A single tear slowly rolls down his cheek. The message is clear. Native Americans and other Indigenous peoples were the original conservationists, and it is modern people who litter and otherwise destroy the natural world.

The crying Native American was a symbol of the wildly popular and successful antilitter campaign by the Keep America Beautiful organization. The advertisement played on (and reinforced) the common misconception that Indigenous people of the past lived in perfect harmony with the world around them. They never wasted resources, they never took more than they needed, and they had the utmost respect for Mother Earth. This image became, and in many circles remains, the quintessential symbol of environmental idealism.

There is no doubt that the human-environmental impacts of the postindustrial world far outweigh those of the past. However, people of the past did alter their environments in considerable ways. Indigenous people modified and tended environments using hunting, gathering, and fire. They moved plants and animals from one place to another to suit their needs, they modified land and seascapes, they diverted rivers and watersheds to access freshwater, and they felled forests to build their homes and monuments and to supply their hearths. Their actions have left an indelible mark on the world today. If we truly wish to understand modern environments and chart effective conservation plans for the future, we must study and incorporate these histories. Modern ecosystems are the product of millennia of human activities, operating in concert with natural systems.

●

Chapter 3

CLIMATE OF CHANGE, NOW AND THEN

As COVID-19 spread across the planet, many of us sheltered in place, drove our cars less, and worked from home. Manufacturing and industry paused, shipping and transportation networks stalled, and life seemed to slow. News outlets reported less smog in our busiest, most populous cities. Stream and river waters were less polluted. Wild animal communities on both land and sea flourished. Earth's ecosystems had a moment to recover and responded quickly. In total, the global economic hiatus of 2020 reduced annual carbon emissions by about 7 percent. In the midst of a dire time, this news was a silver lining. Yet the two gases that contribute most to anthropogenic climate change, carbon dioxide and methane, continued their unrelenting rise.

Carbon dioxide and methane are called greenhouse gases because, like the glass panes of a greenhouse, they trap solar radiation and warm our planet's atmosphere. Other atmospheric gases, such as oxygen and nitrogen, do not have this effect. Despite the pause in human industry, between the beginning of 2020 and April 2021, levels of atmospheric carbon dioxide increased by 12 percent to 43.5 parts per million (ppm). Carbon dioxide levels in our upper atmosphere are now on a par with those of the Mid-Pliocene Warm Period 3.6 million years ago, when concentrations ranged from about 380 to 450 ppm. On April 3, 2021, carbon dioxide concentrations exceeded 420 ppm for the first time in recorded history. Methane, which is less abundant but twenty-eight times more potent as a greenhouse gas than carbon dioxide, also saw surges. The 2020 annual increase in atmospheric methane was 14.7 parts per billion (ppb), the largest since measurements began in 1983. The net result has been a two-thirds of a degree Fahrenheit increase in the world's temperature in just the last twenty years and a 2°F (1°C) increase since the Industrial Revolution. This may not sound like much, but it is an alarmingly fast rate of increase.

You might ask, so what? The Mid-Pliocene Warm Period sounds pleasant enough. At that time, however, average global temperatures were 7°F (3.8°C) higher, Greenland was mostly green, large forests occupied the Arctic, and sea levels were nearly eighty feet higher than they are today. A return to anything close to these conditions would spell disaster for humans. Ocean circulation would be fundamentally disrupted, our agricultural systems would fall into chaos, and coastal cities around the world would be inundated. We would face unprecedented heatwaves, freshwater shortages, massive storms, and the extinction of a million plant and animal species. Human societies would be hard pressed to adapt.

The Global Monitoring Laboratory (GML) of the US National Oceanic and Atmospheric Administration (NOAA) compiles data and conducts research on the long-term levels of atmospheric gases, aerosol particles, clouds, and surface radiation. GML scientists tally fluctuations in atmospheric conditions and attempt to determine their causes and consequences. Based in Boulder, Colorado, the GML maintains four observatories, one each in Hawaiʻi, Alaska, and American Samoa, and one at the South Pole. It also analyzes data collected by volunteers at more than fifty other sites around the world. These data are meticulously recorded and used by climate scientists to understand and predict planetary changes. These researchers agree that the Earth's temperature increase since the Industrial Revolution cannot be explained by natural causes. Humans are to blame. In particular, the burning of fossil fuels such as coal, oil, and gas, on which manufacturing and transportation systems depend, has spewed carbon dioxide, methane, and other greenhouse gases into our atmosphere and triggered these temperature increases. In the words of Colm Sweeney, associate director of science for the GML, "Human activity is driving climate change. If we want to mitigate the worst impacts, it's going to take a deliberate focus on reducing fossil fuel emissions to near zero—and even then, we'll need to look for ways to further remove greenhouse gases from the atmosphere."

Human-induced climate change and global warming are real. Our direct actions and decisions are warming the world's climate at a rate that is unsustainable and will result in dire consequences for life on Earth. Rather than continuing to argue about whether climate change is real and whether humans are the primary cause, we should be discussing whether we should or can do anything to mitigate human impacts on Earth's climate. Some might argue that since the consequences of anthropogenic climate change are likely to be greatest for our children, grandchildren, and great-grandchildren, we can leave it to them to address. Similarly, an argument could be made that we should continue with business as usual because the costs of shifting our fossil-fuel-based economies to more sustainable strategies outweigh the benefits. Others might simply conclude that nothing can be done.

By contrast, scientists and other experts are unanimous in the view that we must immediately address anthropogenic climate change and implement strategies to curb greenhouse gas admissions.

Time is running out. Every eight years the Intergovernmental Panel on Climate Change (IPCC), the United Nations body for assessing climate-change science, releases an assessment report, reviewing thousands of academic publications. The most recent, published in April 2022, ran to nearly three thousand pages. The report was a damning litany of failed promises, missed targets, and inaction by governments around the globe. The IPCC working group chair, Jim Skea, warned that "it's now or never" if we want to avoid dramatic shifts in Earth's life-support system.

Contending with climate change and global warming will require an international effort by citizens, governments, corporations, and scientists. We need to lower atmospheric carbon dioxide and methane levels, reduce our reliance on fossil fuels, and shift to alternative, sustainable sources of energy. In this effort the past can offer crucial insights into how human activities have contributed to global warming and how we might develop mitigation strategies. Although humans have been engaged in the widespread burning of fossil fuels only since the Industrial Revolution began in the 1700s, there is strong evidence that human-generated atmospheric increases of carbon dioxide and methane date back thousands of years. The archaeological record and deep history can expand our perspective and help us devise ways to reduce greenhouse gases and lower their concentrations to more sustainable levels.

EIGHT THOUSAND YEARS OF PLOWS, FORESTS, AND GREENHOUSE GASES

The late 1990s marked a turning point in the scientific study of Earth's greenhouse gases and their linkage to climate change. The scientific community had long been aware of the greenhouse effect.

The Irish physicist John Tyndall is commonly credited with discovering it in 1859, and he was probably beaten to the punch three years earlier by Eunice Foote, a female scientist who conducted experiments by filling different glass jars with water vapor, carbon dioxide, and air and comparing how much they heated up in the sun. But it was not until about thirty years ago that the publication of new, high-resolution data from the ice cores of Greenland and Antarctica enabled wider study of historical changes in atmospheric greenhouse gas levels. Ice cores are cylinders of ice drilled from the interior of ice sheets or glaciers, which are up to three kilometers deep. These cores reveal the distinct layers of ice that form year after year as snow accumulates on the ice sheets. Each year the snowfall has a different texture and chemistry, and winter snow is distinct from summer snow. These deposits build up like the layers of a cake. Examining these layers in the ice cores gives us information about annual wind patterns, rainfall, temperature, and atmospheric composition over long periods of the Earth's history. From Greenland, ice core records extend back 123,000 years, and from Antarctica, they date back 800,000 years. Small bubbles of air trapped in the layers of ice make it possible to measure the historical concentration of greenhouse gases like carbon dioxide, methane, and nitrous oxide with remarkable precision. Scientists have been directly monitoring and recording measurements of greenhouse gas concentrations in the atmosphere only since the 1950s, so ice-core records offer the best way to compare modern conditions with those of the deep past. We are confident in the findings from ice-core records because analysis of air bubbles from ice cores in Antarctica from the 1980s has produced results highly consistent with direct atmospheric measurements.

Starting in the 1990s, the marine geologist William Ruddiman participated in the summer meetings of the Cooperative Holocene

Mapping Project, an interdisciplinary group of scientists interested in understanding Earth's changing climatic conditions from
the end of the last Ice Age, around twenty thousand years ago, to
the present. The team noticed that concentrations of atmospheric
greenhouse gases did not match the patterns predicted by the cyclical climatic oscillations postulated by Milutin Milanković (see chapter 1). Ruddiman found that concentrations in carbon dioxide started
to rise about eight thousand years ago, at a time when, according to
assumptions based on the Milanković cycles, they should have been
declining. About five thousand years ago, concentrations of methane
also rose when they should have declined. Ice-core samples about six
thousand to four thousand years old, for example, register a nearly
tenfold increase in methane compared to previous millennia. Without
these increases, northern parts of North America and Europe would
have been three to four degrees Celsius cooler than they are today,
and a mini Ice Age in northern latitudes would have begun several
thousand years ago. The net result was a relatively warm and stable
climate in the past few millennia prior to the Industrial Revolution.

Most paleoclimatologists pointed to natural factors to explain
Ruddiman's findings. Carbon dioxide increases were explained by
a combination of the natural loss of carbon-rich vegetation on the
continents and geochemically driven changes in ocean dynamics,
such as ocean circulation, temperature, and productivity. Methane
increases were tied to the expansion of methane-emitting wetlands
in the Arctic. The major flaw in these explanations is that virtually
the same natural conditions were in place over the four preceding
interglacial periods. Why would atmospheric carbon dioxide and
methane concentrations have fallen during the last four interglacial
cycles but risen during the current one? For Ruddiman, the only
reasonable answer was the presence of humans.

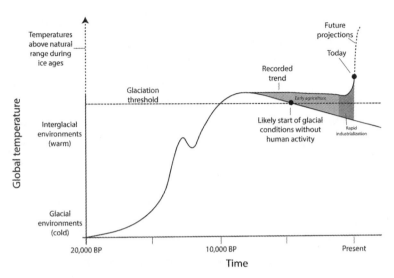

Average global temperatures since the Last Glacial Maximum. The shaded area shows the difference between observed global temperatures and the predicted temperature trend without anthropogenic greenhouse gas emissions by early agriculturalists. Work by William Ruddiman shows that early agricultural activities and deforestation starting eight thousand years ago accelerated global warming trends long before the Industrial Revolution. Coincident with industrial-scale human activities, global temperatures spiked. Projections suggest that this trend will continue.

Ruddiman spent much of his subsequent academic career developing this hypothesis. In 2005 he published his landmark book *Plows, Plagues, and Petroleum: How Humans Took Control of Climate,* on the history of human agricultural activities and global climatic patterns. In what has become known as the Ruddiman hypothesis, he argued that humans have been altering global climate since the advent of widespread agriculture eight thousand years ago. Through farming and forest clearance, Ruddiman suggested, human activities kept our planet warmer than natural climatic cycles would have dictated and quite possibly forestalled an ice age event.

63

Ruddiman's interpretations sparked vigorous debates. A number of climate scientists, including the prominent geochemist Wally Broecker from Columbia University, known as "the grandfather of climate science," held firm on natural explanations for these early greenhouse-gas patterns. Prior to about the nineteenth century, these researchers argued, human populations were not large enough to clear tracts of forest big enough to alter global carbon dioxide concentrations. Although further work by Ruddiman and others shows strong links between human activities and increases in greenhouse gases, some geochemists and geoscientists still maintain that any early human effect on global climate was a small or transitory one.

One of Ruddiman's most powerful arguments hinges on the differences between the most recent interglacial period and the earlier ones. By the last interglacial, humans had spread out of Africa and across much of the globe. The transition to agricultural economies was well under way in the eastern Mediterranean (the Fertile Crescent region) and northern China at least eleven thousand years ago. With the advent of agricultural food systems came changes in human settlement patterns and population increases. According to Ruddiman, the primary drivers of carbon dioxide and methane increases starting eight thousand years ago were the transition from hunting and gathering to pastoralism and agriculture, which led to a dramatic increase in human populations, and the shift from small, mobile societies to large, urbanized states.

Human activities can generate greenhouse gases even without burning fossil fuels. To expand farming plots or grazing land, humans clear forested landscapes. Trees store a considerable amount of carbon, and, when forests are burned or felled and left to rot, this stored carbon is released into the atmosphere as carbon dioxide. By eight thousand years ago, early farming communities in

Europe and China were clearing vast tracts of previously forested landscapes, unwittingly releasing large amounts of carbon dioxide. As farming communities grew, animals were more widely domesticated, and higher agricultural yields were required to feed more and more people, this process transformed landscapes around the globe. In 2018, a group of scientists used pollen data combined with computational modeling to estimate the past size of forests. They concluded that forest cover peaked in Europe about eight thousand years ago. By three thousand years ago, deforestation had transformed the European landscape from a largely forested one to an agricultural mosaic—a mixture of settlements, agriculture, forests, and other land uses.

This research was followed by a study conducted by a consortium of over 250 archaeologists and historical scientists, who concluded that by at least three thousand years ago, human activities had transformed the face of the planet, primarily through the creation of pastoral and agricultural land. The idea that early human societies were not large or sophisticated enough to alter landscapes significantly is simply not accurate.

Atmospheric methane levels began to increase about three thousand years after carbon dioxide increases had begun, when agricultural practices were established throughout much of Europe and Asia. A number of factors likely contributed, including increases in herds of livestock, whose feces and belches emit methane gas, and the burning of grasslands as a landscape-management strategy. The primary driver, however, was likely rice cultivation in southern Asia. Beginning about five thousand years ago, farmers in southern China took to flooding lowlands along river valleys to grow rice. These artificial wetlands release methane in much the same way natural wetlands do. Flooding traps vegetation beneath stagnant water. As the vegetation decomposes, it releases methane into the atmosphere.

Ruddiman and a team of Chinese scientists found a strong correlation between increases in wet-adapted rice agriculture throughout China and increases in atmospheric methane beginning about
five thousand years ago. Archaeobotanical data (plant remains preserved in archaeological deposits) have been used to track the spread
of rice-paddy agriculture throughout the region, and these findings
lend support to Ruddiman's conclusions. By three thousand years
ago, farming communities in Vietnam, Laos, Cambodia, and India's
Ganges River Valley had adopted this practice. By two thousand
years ago, agricultural intensification drove farmers to transform
steep hillsides in Southeast Asia into rice paddies, contributing to
the steady rise of methane concentrations.

Through the invention of domestication and agriculture and
the deforestation and landscape modifications that followed, humans
helped create a warmer global climate. Because this was more suitable for human agrarian systems, in a sense humans reshaped the
global climate to their advantage. By the dawn of the Industrial
Revolution, these changes had led to increases of 40 ppm in atmospheric carbon dioxide and 250 ppb for methane. This would place
preindustrial human-induced warming at almost 0.8°C, on par
with or slightly more than anthropogenic warming over the past
century.

THE LITTLE ICE AGE AND DISEASE

Although these human effects on the planet's atmosphere and climate mostly predate recorded history, we have one example within
recorded (written) human history that offers direct evidence of temperature fluctuations and localized changes. Even with written
descriptions of the changes, however, explaining the causes of these
climatic shifts is not easy.

Beginning in the early fourteenth century CE, Europe experienced what has been dubbed the Little Ice Age. This is not the most accurate moniker, as ice age events were global and a hallmark of the Pleistocene—the time when mammoths, mastodons, and other cold-weather-adapted giants roamed the planet. The Little Ice Age was a regional phenomenon mostly affecting the North Atlantic, Greenland, and northern Europe. A better name might be the Little European Ice Age. During this period, average global temperatures dropped as much as 2°C (3.6°F), but there was tremendous variability: it was not always cold, and it was not cold everywhere. Regardless, the chilling effect caused the expansion of glaciers in Scandinavia, the Alps, Iceland, Alaska, China, the Andes, and New Zealand. Londoners laced up their ice skates and glided down the River Thames. Birds froze to death in midair and fell from the sky; people died of hypothermia. Wet summers and icy winters resulted in reduced crop yields and famine in northern and central Europe. A sixteenth-century Icelandic poem describes the suffering many felt:

> Formerly the earth produced all sorts
> of fruit, plants and roots.
> But now almost nothing grows . . .
> Then the floods, the lakes and the blue waves
> Brought abundant fish.
> But now hardly one can be seen.
> The misery increases more . . .
> Frost and cold torment people
> The good years are rare.

The Little Ice Age can be divided into two phases: the first cooling phase, which began in about 1300, and a second, harsher and more abrupt, which started around 1570 and lasted until 1680.

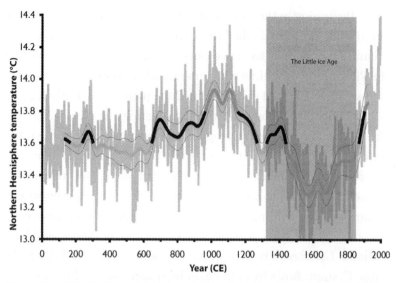

Average temperatures in the Northern Hemisphere over the last two thousand years. Highlighted is the Little Ice Age, when temperatures dropped by as much as 2°C. (Base graph image via Wikimedia Commons.)

The start of the Little Ice Age likely had nothing to do with human activity. It was probably triggered by a large eruption at the Samalas volcano on the Indonesian island of Lombok in 1257, followed by three smaller eruptions over the next four decades. This volcanic activity sent large volumes of aerosol particulates into the atmosphere, which reflected more of the Sun's rays back into space and cooled the Earth. This cooling increased northern sea ice, which in turn kept temperatures cooler for longer periods, creating a feedback cycle.

For our purposes, the second phase of the Little Ice Age is much more interesting. In 1492, human activity changed the world in fundamental ways. The arrival of Christopher Columbus in the Americas ushered in the era of European colonization. Columbus and his fellow European explorers had little idea of the majesty and

68

complexity of the world they sailed into. One of the common misconceptions about Native Americans has been that they lived predominantly in small groups of nomadic hunter-gatherers. This was true in some parts of the Americas, especially in resource-poor regions, where people employed ingenious adaptations to survive and thrive. But large and complex civilizations were thriving in North, Central, and South America long before European arrival. The Inca Empire, for example, stretched across western South America from Quito (in modern Ecuador) in the north to Santiago (in Chile) in the south. Tenochtitlan, the capital of the Aztec Empire, located in what is now the center of Mexico City, was among the largest cities in the world, with some two hundred thousand residents. Not long before Columbus landed, the agricultural city of Cahokia, near modern-day St. Louis, was the largest of its kind in North America, with over twenty thousand inhabitants.

While estimating pre-Columbian populations is notoriously difficult, a recent study compiled multiple sources of evidence to conclude that there were some 60.5 million people living in the Americas in 1492. This number is smaller than populations in Europe at the beginning of the sixteenth century (estimated at 70–88 million) or in China and Mongolia (100 million), but it is still a sizable human population. Feeding all these people required large tracts of land for agricultural fields. In Inca territory, steep hillsides were terraced to grow maize, quinoa, and other staple foods. Fire and irrigation were important tools for managing and maintaining agricultural fields. In Mexico, vast tracts of land were cultivated for maize, cacao and fruit orchards, and house gardens. Elaborate canal systems were constructed along the Gulf Coast in Mexico, and slash-and-burn agriculture was important in the Yucatán. Mississippian people in the Midwest and southeastern United States relied on maize, squash, and beans grown in alluvial areas. These groups also constructed massive earthen mounds as living platforms and centers of ritual activities. Over millennia, Indigenous Americans had developed extensive and intensive land-use strategies to feed populations numbering in the tens of millions. The effect of their combined efforts was a heavily modified and deforested New World landscape.

When Columbus and other European explorers arrived, they brought with them infectious diseases to which Native Americans had little to no immunological resistance. Viruses rapidly swept through Indigenous populations, from the points of contact in coastal regions to areas far inland, as Native communities traded and interacted with one another. Deadly waves of smallpox, measles, whooping cough, chicken pox, bubonic plague, typhus, and malaria reached inland communities well in advance of the first European explorers. Disease, followed by war, slavery, and the dehumanizing effects of European occupation, wiped out entire communities.

The most recent studies estimate that European arrival likely resulted in fifty-five million Native American deaths by 1600. To put this in perspective, during the COVID-19 pandemic, global deaths exceeded five million by late 2021, causing deep suffering and loss. The pandemic has left an indelible mark on an entire generation, and our lives may never be the same. Yet global deaths from COVID stand at just a fraction of a percent of the world's eight billion people. The impacts of disease epidemics and colonization on Native American communities during the sixteenth century were on a different scale altogether.

As populations crashed, Indigenous cities, communities, and sociopolitical and economic networks disintegrated. Vast tracts of farmland were abandoned as the demand for food declined, along with the labor available to produce it. The ecological succession process reclaimed cleared land. Fifty-six million hectares (two hun-

dred thousand square miles) of agricultural fields reverted to forest during the sixteenth century, an area roughly the size of Texas,
or 10 percent of the Amazonian rainforest. Because growing plants
absorb carbon dioxide, the extra vegetation pulled an enormous
amount of carbon dioxide from the atmosphere (estimated at about
5 ppm). These human-initiated events likely triggered the second,
more abrupt cooling event of the Little Ice Age by sequestering
carbon dioxide in the regrowth of New World forests and altering
global atmospheric conditions.

These events and their consequences offer powerful examples
of how human action affected global climatic conditions long before
the Industrial Revolution. They also suggest that effective land-
management strategies and the regrowth of forests may hold large
potential for mitigating climate warming and meeting global climate and sustainability goals.

CLIMATE TODAY AND INTO THE FUTURE

When we think about anthropogenic climate change and greenhouse warming, it is easy to fall into the trap of hopelessness. Fossil-
fuel use seems so firmly entrenched in the global economy that it
is hard to see an alternative. But there are a number of reasons to
remain optimistic about the future. We have all the technologies we
need to dramatically reduce our reliance on fossil fuels and reverse
the trend of increasing atmospheric carbon dioxide and methane.
Attitudes towards these green technologies have undergone a sea
change in recent years. Who would have guessed ten years ago
that energy-efficient electric cars would become a status symbol
in American culture? Over the past decade, the Tesla has replaced
the BMW as the "It" car for signaling status and prestige among
southern Californian middle-class families. These attitude shifts

may help us achieve what we need and need quickly: lower energy consumption, continued electrification of transport systems, and enhanced carbon storage (that is, removing carbon dioxide from the atmosphere).

The events of past millennia and the Little Ice Age demonstrate that deforestation contributes to rising levels of carbon dioxide and methane, and that reestablishment of large forests can contribute to global cooling by reducing atmospheric carbon dioxide. These phenomena can help us predict future greenhouse-gas emissions and establish baselines by which to set mitigation targets.

The socioecologist Karl-Heinz Erb at the University of Vienna and a number of colleagues published a paper on the potential impacts of forest and animal-grazing management on global vegetation. They argued that well-planned land management would make it possible to naturally capture and sequester huge amounts of carbon from the atmosphere, enabling us to quickly progress toward meeting the goal of the 2015 Paris Agreement on climate change to limit global warming to a 1.5°C increase above the preindustrial temperature. Reforesting areas of our planet could very quickly build biomass, essentially removing massive loads of carbon dioxide from the atmosphere and mitigating greenhouse warming.

What this all boils down to is that we need not rely on some yet-to-be developed carbon-capture technology to save us. Nature can provide. Large tracts returned to forested states could become huge stores of carbon. The trick will be to rethink our land-management strategies, find ways to preserve and protect the world's forests, and support political, social, and economic leaders who share this vision.

○

Chapter 4

NO TREES FOR NOTRE-DAME?

When I was young, I was lucky enough to visit Paris twice: once when I was a naive high school junior on a French class trip and again as a slightly less naive college graduate on a backpacking adventure through western Europe. To someone who grew up in small-town Indiana, the streets of Paris seemed a different world. What captured my imagination was not the big-city hustle and bustle (Chicago was only an hour away from my home) but the history, elegance, and physical beauty of Paris.

On my first visit, I imagined myself a struggling artist haunting the cafés and bars along the Left Bank, sharing stories with expatriate authors like Gertrude Stein and Ernest Hemingway, using Paris as my muse while I composed the next great American novel. Yet the city exceeded even my romantic expectations. I returned five years later, broke but still starry-eyed, with very little sense of direction for my life. Paris gave me a place to reflect on the world and how I might find my place in it. I viewed the Mona Lisa. I sat at the base of the Eiffel Tower. And I stared in awe at the stained-glass windows in the cathedral of Notre-Dame de Paris.

Photograph taken on April 15, 2019, as Notre-Dame's iconic spire burned and was destroyed in the early evening. (Image via Wikimedia Commons.)

At 6:20 p.m. on April 14, 2019, under clear blue skies, fire alarms rang out from the Île de la Cité. The almond-shaped island is the heart of Paris, the site of the original medieval city, and the home to some of the modern city's most important civic and cultural institutions. At its eastern end is Notre-Dame. For 850 years, the Catholic cathedral has awed tourists and worshippers with its massive flying buttresses, stained-glass windows, intricately carved sculptures, colossal church bells, and holy relics of Christendom. It was the site of Napoleon I's coronation in 1804, the setting for Victor Hugo's 1831 *The Hunchback of Notre-Dame*, and the heart of celebrations after the liberation of Paris from Nazi control in 1944. Now the cathedral was on fire.

For twelve hours smoke billowed from the cathedral, and more than four hundred firefighters battled the inferno. When the last embers were finally extinguished, most of Notre-Dame's wood and metal roof was destroyed, and its majestic spire had been consumed by the flames. A visibly emotional President Emmanuel Macron mourned, "It is a terrible tragedy. . . . Notre-Dame is the cathedral of all French . . . the epicenter of our life."

The global outpouring of support, both moral and financial, was swift and spectacular. Political and religious leaders around the world offered their condolences; wealthy philanthropists, international companies, and blue-collar workers opened their checkbooks to contribute to the cathedral's restoration. Within days, more than a billion dollars had been offered to the French government. Two days after the fire, however, CNN reported an unexpected complication. The massive beams that had framed the roof were fashioned from locally harvested European beech trees that had been three hundred to four hundred years old when they were felled. European beech is one of the most common hardwood trees throughout north and central Europe, but the vice president of the French Heritage Foundation, Bertrand de Feydeau, lamented to CNN, "The roof was made of beechwood beams over eight hundred years ago. There are no longer trees of that size in France." Asked if trees of adequate size could be found outside France, he answered, "I don't know."

Other tree species could be considered, or engineered materials might offer functionally acceptable replacements for the beechwood beams, but either option would compromise the historical integrity of Notre-Dame's architecture. Unless you hold a keen interest in historic preservation, this might seem like a small problem. But in fact it highlights an environmental threat of global significance: the vast clearance of forests and woodlands for cropland, pasture, fuel wood, and construction materials.

The problems related to deforestation and landscape clearance have been building in scale and intensity for millennia. Rich paleoecological and archaeological records in Europe suggest that intensive, continuous human occupations by agriculturalists likely triggered successive cycles of deforestation, abandonment, and afforestation (the planting of crops on land barren of trees, such as old agricultural fields) for at least six thousand years. More than fifty years ago, H. C. Darby, a British pioneer in historical geography, argued that deforestation was "probably the most important single factor that has changed the European landscape." To grasp the scale of the problem and consider how we might effectively combat it, we need to begin at the dawn of the Agricultural Revolution in Europe, when the introduction of a variety of nonnative plants and animals initiated a cascade of ecological, social, and political transformations.

THE ORIGINS OF AGRICULTURE IN EUROPE

At the height of the Pleistocene geologic epoch (or the Last Glacial Maximum, between twenty-two thousand and nineteen thousand years ago), commonly referred to as the Ice Ages, Europe was a strange and hostile place. Temperatures were on the order of 20°C (36°F) colder than the present, and the northern reaches of the continent were blanketed in an ice sheet over 4 kilometers (2.5 miles) thick. With so much water trapped in polar and high-altitude ice, the world was arid. Sea levels dropped 100 meters (328 feet) below their present positions, exposing vast tracts of continental shelf. Britain was connected to the European mainland. A list of cold-adapted creatures roamed the continent like the cast of a science fiction movie, including woolly mammoths standing 3.4 meters (11 feet) tall and weighing six metric tons (13,200 pounds), woolly rhinoceroses, and

Irish elk with the largest antlers of any deer ever known, extending 3.65 meters (12 feet) from tip to tip. Lurking to prey on these creatures were nightmarish carnivores, including cave lions 25 percent larger than today's largest African lions and cave bears weighing 500 kilograms (1,100 pounds), along with cave hyenas, saber-tooth (or scimitar-tooth) cats, and giant polar bears.

By about eighteen thousand years ago, the world was steadily warming. The transition to the warmer Holocene was anything but smooth, however. For about two millennia beginning 14,700 years ago, the Bølling-Allerød interstadial period brought warm and moist conditions across much of the globe, with summer temperatures reaching nearly their present levels. The rapid melting of glacial ice released massive quantities of freshwater into the North Atlantic, slowing the ocean conveyor belt (thermohaline circulation), which, paradoxically, helped trigger a 1,300-year cold spell (the Younger Dryas) and a return to Ice Age conditions until about 11,500 years ago. At the end of the Younger Dryas, rapid climatic warming resumed. Global temperatures rose by about 7°C (12.6°F) within half a century, ushering in the Holocene epoch. In short order, many of the giant animal species of the late Pleistocene world died out.

Ecosystems in Europe responded to the increased temperatures and rainfall of the early Holocene in dramatic ways. Cold-climate adapted trees of the northern boreal forest, such as birch and pine, were pushed northward as ice sheets melted and retreated. Over several centuries, deciduous trees such as oak, elm, and beech took their place, slowly spreading out from Ice Age refugia that contained the developed soils capable of supporting them. Across Europe and in northern latitudes around the globe, open tundra gave way to thick forest, and Ice Age megafauna species were replaced by wild boar, red deer, and other smaller game. For humans, the warmer and wetter conditions offered tremendous opportunities for hunter-gatherer communities in the newly temperate zones of Europe.

During the harsh conditions of the Pleistocene in Europe, human populations were concentrated in tropical and subtropical zones as small, nomadic groups. As warming increased the overall biomass of the planet, hunter-gatherer communities expanded out from central latitudes and settled along forest breaks, lakes and rivers, and wetland margins. They hunted game with bows and arrows, speared fish during seasonal runs, and harvested wild tubers with digging sticks. Geographic and seasonal variations in the availability of resources forced small-scale communities to move their settlements regularly.

As the Ice Age drew to a close, a small number of humans began to experiment with new ways to exploit their environments. Rather than simply collecting plants and seeds, some groups intentionally managed plant growth adjacent to their settlements, making food plants more readily available and reducing the time spent foraging. These Stone Age innovators quickly discovered that seeds could be saved and planted for another year's crop. They also discovered that seeds with certain traits stored better, grew better, or produced superior grains. Year over year, their intentional selection of plants with desirable traits resulted in changes to the genetic composition of the plants they sowed and harvested. The continued manipulation of wild plants resulted in species that could not survive without human assistance: the intentional preparation of fields and the sowing, harvesting, and storing of seeds. The biological process of domestication established a symbiotic relationship between humans and their domesticates. The clearing of forests, harvesting and storage of crops, and development of new farming technologies led to large permanent villages, growing population densities, and increasingly complex sociopolitical organization.

One of the first areas where this transition was made was in the Fertile Crescent, a quarter-moon-shaped section of the Middle East stretching from the Persian Gulf through modern-day southern Iraq, Syria, Lebanon, Jordan, Israel, and northern Egypt. By the end of the Ice Age, the occupants of this region had congregated into villages. For food they hunted and gathered boar, deer, gazelle, fish, and mollusks. Wild cereals (plants that produce starchy grains), including wild barley and emmer wheat, also constituted a significant portion of the diet. All grew abundantly in the rich soils of the Tigris and Euphrates River valleys, the Mediterranean woodland zone, and the larger Fertile Crescent region. This virtual Garden of Eden allowed hunter-gatherers to build permanent communities, process foraged grain, and store their surpluses.

The return of colder and drier conditions during the Younger Dryas period exerted tremendous stress on these communities. As wild cereal crops decreased in abundance, people began selecting, planting, and tending large, hardy seeds selected from their gathering. By eleven thousand to ten thousand years ago, new forms of barley, along with emmer and einkorn wheat, emerged in the Fertile Crescent, distinct in size and shape from their wild cousins. Also domesticated were legumes such as lentils and peas.

Around this time, people also were experimenting with the domestication of animals. Archaeologists have unearthed evidence for the herding of goats, sheep, and cattle in the Fertile Crescent at least ten thousand years ago. Pastoralists controlled the population profiles of their herds, retaining females and juveniles and consuming subadult males, and manipulated their genetic structure by selecting for desirable traits such as docile dispositions. This investment in animal husbandry resulted in dramatic and rapid shifts in human diets. The University of Arizona archaeologist Mary Stiner and her team of researchers, for example, found that at the Aşıklı

Höyük site in central Turkey, communities shifted from a hunting and gathering lifeway to a nearly exclusive reliance on sheep for dietary protein over only two centuries. Mobile hunter-gatherer groups became sedentary farming communities. The Agricultural Revolution had begun.

Pinning down where and when crops and animals were first domesticated is a tricky business, because the archaeological remnants of such transitions are murky. Technologies such as genetic sequencing, isotopic studies, residue analyses, high-precision dating techniques, and increasingly meticulous excavations are helping scientists make rich discoveries. Surprisingly, Europe, with all its ecological diversity and deep human history, is the one continent (save Antarctica) where no economically important crops or animals were independently domesticated, except perhaps oats and some legumes. Virtually every plant and animal significant in the story of agricultural transitions in Europe was introduced from the south and east. Einkorn wheat, barley, beans, lentils, cows, sheep, and goats all originated in the Fertile Crescent or elsewhere in the Near East. There is no evidence for wild progenitors of these plants and animals in Europe: when they first appear in the archaeological record, they do so in their fully domesticated form. Archaeological excavations reveal that the transition to agriculture began in Europe about nine thousand years ago, resulting in the so-called Neolithic (New Stone Age) revolution.

The question that has long plagued archaeologists is how these domesticates ended up in Europe. Clearly they did not get there by accident. Were they introduced through a succession of trades among neighboring human groups, or by migration of new human groups into the area? Genetic studies suggest that the mechanism was likely migration. A comparison of DNA extracted from human skeletal material from some of the first farmers in central Germany with the

DNA of modern groups, for example, revealed that Neolithic farming communities in the area were likely established by migrants from the Near East and Anatolia. Another study sequenced the genomes of a farmer from Germany who lived seven thousand years ago, eight hunter-gatherers who lived in Luxembourg and Sweden eight thousand years ago, and over two thousand modern Europeans. The study showed a significant contribution of genetic ancestry from Middle Eastern agriculturalists, who arrived in European landscapes about nine thousand years ago. Genetic analysis of ancient and modern populations reveals a trail of unbroken ancestry and human migrations from Europe back to southwest Asia and perhaps to the Fertile Crescent.

Once domesticated crops reached Europe, the transition from hunter-gatherer to agricultural economies was a relatively rapid one, although its spread was uneven. One sign of agricultural communities is the creation of pottery and ceramic vessels; more mobile hunter-gatherers tended to use lighter containers made from perishable materials like gourds and plant fibers. One of the most recognizable early agricultural groups in Europe has been dubbed the Linearbandkeramik (LBK) culture, named after a distinctive type of pottery with linear, incised designs decorating the outside. LBK pottery, consisting of simple cups, bowls, vases, and jugs, conveniently marks the spread of food production into central Europe. Along with pottery, archaeologists have unearthed a variety of agricultural tools at LBK settlements, including sickles, which were crafted by inserting razor-sharp flint blades along the interior margin of a curved wooden hand tool in order to harvest grains, and stone adzes (similar to axes) used to fell trees and shape wood.

From its origins in Hungary, LBK culture spread rapidly across central Europe north of the Alps. European loess is a wind-blown

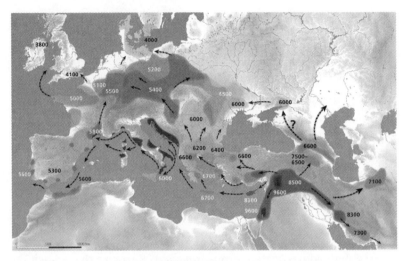

Map showing the spread of early Neolithic farming communities in Europe. Dates are BCE approximations. Shaded areas represent the range of the farming communities in that time period. (Image via Wikimedia Commons.)

sediment uncovered by the retreat of glaciers during the Pleistocene, which, under certain environmental conditions, develops into very rich soils and can accumulate in alluvial environments. A massive loess belt stretches across northern Europe, from France through Germany and Poland to northern Ukraine. LBK cultures found loess accumulations along major river systems and quickly established settlements to take advantage of the nutrient-rich soils and freshwater. Often, however, these desirable agricultural sites were covered by dense forest. To make way for crops, trees were chopped down with flint and polished-stone axes. The densest archaeological evidence for LBK cultures has been found along three of the largest river systems in Europe: the middle Danube, the upper and middle Elbe, and the upper and middle Rhine. LBK cultures established relatively small settlements with agricultural plots along 1,500 kilometers (about 930 miles) of riverine terraces in less than four hundred years.

LBK cultures built massive longhouses out of wattle and daub (woven wooden strips coated with a sticky combination of wet soil, clay, sand, animal dung, and straw). These dwellings were 5.5–7 meters (18–23 feet) wide and up to 45 meters (148 feet) in length. They were divided into work and sleeping areas, with pens for domesticated animals. Villages included a handful of longhouses, each supporting an extended family, distributed over about four hundred hectares (one thousand acres). Farmers planted small fields (often referred to by archaeologists as *gardens*) of cereal grains and legumes, and they pastured livestock along forest margins. Cattle supplied meat, milk, and cheese. Sheep and pigs also provided protein and other resources.

As farming settlements expanded, they came into contact with local hunter-gatherer groups, who were slowly absorbed into agricultural communities. Like ripples on a pond, farming lifeways spread across Europe over the next two thousand years. Slash-and-burn cultivation and other forms of forest clearance transformed the continent. Natural landscapes became anthropogenic ones.

With agriculture came all the hallmarks of sociopolitical complexity. Settlements grew, agricultural production increased, religious and political leaders emerged to organize labor, and artisans created high-value goods. Cities were established, irrigation canals were constructed, and awe-inspiring monuments like Stonehenge were erected. All these changes have left enduring impacts on local and regional environments. But hidden in plain sight is the most influential and enduring legacy of European agriculture: the clearance of forests and woodlands for cropland, pasture, fuel, and construction materials.

RETURN TO PARIS

Twenty-three years after my last visit, I boarded a plane for Paris in October 2021. Even in the middle of the COVID-19 pandemic, I longed to see the injured cathedral of Notre-Dame, and I desperately wanted to pass some time sitting at a Parisian café.

By a stroke of good fortune, I was invited to attend a small academic conference on the archaeology of coastlines at the Musée du Quai Branly, a spectacularly beautiful museum dedicated to the Indigenous art and cultures of Africa, Asia, Oceania, and the Americas. With my wife's blessing, I extended my stay by a couple of days to visit Notre-Dame and some of the other sights.

In Paris, I attended my conference at Quai Branly and learned about some of the amazing archaeological projects being carried out along the French coast and beyond. I gazed at priceless artifacts and ethnographic pieces from around the globe that tell the history of human innovation, ingenuity, and cultural diversity. I caught up with colleagues new and old at random cafés while seated under red awnings, neon signs, and the shadow of the Eiffel Tower. But my real goal, to visit Notre-Dame and survey the damage, had to wait until my final days in Paris. Until then, I could only wonder whether a faithful repair of the cathedral could be accomplished after millennia of deforestation across France and throughout Europe.

EUROPEAN DEFORESTATION: FROM THE
ICE AGE TO GLOBALIZATION

Studying vegetation communities over a large area is arduous work. The larger the study area, the more complicated the task. Botanists employ surveys (walking the landscape and making note of the plants they encounter), aerial photography, and satellite imagery to create models and maps of contemporary vegetation communities and the changes they undergo. Reconstructing the history of vegetation communities is even more challenging and perhaps even more important for conservation science. Long-term vegetation histories help establish the baselines scientists need in order to understand why vegetation communities look the way they do today.

More than half the land on the Earth's surface consists of what environmental scientists call *anthromes*: landscapes that are defined not only by their climate, topography, hydrography, and vegetation, but also by human activities such as urban development, farming, irrigation, and rangeland construction. Over generations, these human activities transform landscapes into mosaics of varied uses. Anthromes are best thought of as human systems with natural systems embedded within them.

Glance outside. Whether you see a manicured lawn, a community park, your neighbor's apartment complex, a cornfield, or a wooded, public easement, how do you know what that space looked like twenty, fifty, one hundred, or one thousand years ago? You might be able to sift through historical photographs of the area and get some answers. Satellite monitoring of global vegetation change is another source of information, but data collection began only in the mid-1970s. You could head down to the public library and find historical maps going back fifty years to a century. Looking back hundreds or thousands of years is a much more difficult task. The best method scientists currently have at their disposal is the study of ancient pollen, a highly specialized field called palynology.

Flowering plants release pollen spores, the sperm-carrying reproductive bodies, into the environment. For human allergy sufferers, the result may be wheezing, sneezing, and sleepless nights. Barely visible to the naked eye, each pollen grain has a unique shape, depending on the species of plant that produced it, and its walls are made from a chemically robust and stable substance, *sporopollenin*. Pollen grains are washed or blown across the landscape. Some fall into lakes, ponds, or oceans, where they settle to the bottom in mud and sediment layers and remain preserved for thousands of years. These stratified deposits serve as time capsules. Environmental scientists date the sediments, recover the preserved pollen, and painstakingly identify and count the grains, from which they can reconstruct the vegetation cover of an area over thousands and millions of years.

Pollen data are used to construct computer-generated models of the abundance of plant species in a specific region in different periods. Such models have been built, calibrated using contemporary monitoring stations, and rigorously tested for a number of regions, including North America and Europe, to show vegetation cover over the last ten thousand years. They can be used to track changes, including land clearance for the creation of agricultural fields.

A recent project by the University of Plymouth environmental geographer Neil Roberts and colleagues investigated European forest cover over the past ten thousand years. They gathered data from over eight hundred sites across Europe and generated land-cover maps at two-hundred-year intervals. They determined that forest cover was at its maximum about eight thousand years ago and has been gradually declining ever since. This pattern is generally consistent with what we know about the cultural and environmental history of Europe.

During the cold and harsh conditions of the Ice Age, much of northern Europe was a temperate grassland environment with only patchy forest cover. As the Eurasian ice sheet melted and retreated (resulting in the extinction of mammoths, mastodons, and other megaherbivores whose grazing helped maintain grasslands), forests gradually spread across the continent. Between about eight thousand and six thousand years ago, this trend reversed. In some regions this reversal preceded the arrival of agricultural communities and may be linked to either the burning of forests by hunter-gatherer groups or natural climatic changes. In other regions, there is compelling evidence that dramatic landscape changes occurred following the arrival of Neolithic communities and the conversion of forests into agricultural fields. In a few places, like southern Germany, the arrival of agriculturalists corresponds with an increase in secondary woodland (forest cover that has regrown in previously deforested areas), but no evidence for an overall decline.

With some regional variation, the rate of forest loss began increasing about four thousand years ago. Thick charcoal layers, suggesting intentional human burning of forest cover to clear agricultural plots, and decreases in forest pollens, followed by increases in cereal grain and weed pollens, show the role of humans in the transformation of European landscapes. The increase in cereal grains shows the spread of agricultural fields and the felling of forests, as does the increase in weeds, which would have thrived in human-cleared landscapes. By three thousand years ago, prior to the introduction of iron tools, Europe had lost approximately one-fifth of its temperate deciduous forests. Mosaic anthromes had been established across the continent. The clearance of old-growth forests created space and opportunities for a variety of early-succession forest species such as fir, birch, spruce, and Mediterranean pine.

After about three thousand years ago, we have additional sources of data about changes to forest vegetation across Europe. During the Hellenistic and Roman periods, philosophers, naturalists, and historians recorded their observations and thoughts about the world around them. The writings of Homer, Strabo, Cicero,

Sophocles, and many other authors recognize humanity's unique ability to alter the natural world. Historical writings detail the loss of European forests to farming, grazing, domestic fuel gathering, shipbuilding, and metal smelting. The Roman author Lucretius, for example, commented that "day by day they [farmers] would constrain the woods more and more to reside up the mountains" of southern Europe until only the most remote and inaccessible slopes remained forested. Documents record the wholesale devastation of "thickly covered . . . wild and untrodden" forests as the Roman Empire expanded across the continent. Timber was critical for smelting new iron tools, for fuel in domestic hearths and community bathhouses, and for provisioning trade, war, and imperial expansion. The increasing need for new agricultural fields forced farmers to engineer and terrace steep slopes.

Lowlands were cleared before highlands, and forests adjacent to rivers were felled before outlying and upland areas. As newly urbanized communities increased demand not only for cereal grains but also for meat and wool, grazing by domesticated cattle, pigs, and sheep helped make deforestation permanent. Overgrazing and forest clearance led to soil erosion: topsoils were swept into rivers and estuaries from places like the uplands of Greece, redepositing sediments, silts, and gravels along human-transformed coastlines. Deforestation created microclimatic changes: some classical texts note warmer temperatures. Still, some wooded areas remained in peripheral locations; some were protected as sanctuaries.

The collapse of the Roman Empire ushered in the Middle Ages, a time of transition for Europeans. Prior to the sacking of Rome and the dissolution of the political and cultural stronghold of the Roman Empire, pagan beliefs and local religious traditions had bestowed special significance on every tree, hill, and body of water and saved some areas or slowed their felling. The guardian spirits of these places had to be placated before any action that would alter them. When Christian ideals displaced these beliefs, new conceptions of the environment took hold. The Old Testament conveys a view of the world that privileges humans above all other forms of creation. In Judeo-Christian thought, humans were meant to be fruitful and multiply, to exploit and tame nature. These ideals conceptually removed humans from nature and represented natural resources as commodities to be exploited to meet human needs. At the same time, the Bible called on people to be caretakers of the natural world. This casting of humans as both the masters over and the stewards of nature marks the origin of the modern environmental dilemma in Western culture.

The fall of the Roman Empire and the political fragmentation of Europe and the Mediterranean led to a temporary decrease in human population densities and the scale and intensity of farming. Major land-transformation projects such as terracing, forest clearance, and wetland reclamation largely ceased. These trends were accompanied by the regrowth of secondary forests in many regions of Europe. The recovery was short-lived, however. Populations surged over the next seven hundred years, increasing fourfold across Europe by the early thirteenth century. There were some eighteen million people living in Europe in 600 CE. That number doubled by 1000 CE and doubled again to 75.5 million in the early thirteenth century. During the initial years of this population explosion, lightly settled or unoccupied areas of northern Italy, Spain, France, western Germany, Belgium, the Netherlands, Luxembourg, and southeast England were transformed into productive agricultural lands by forest clearance and the draining of wetlands. Lands outside the European heartland were systematically integrated into medieval agrarian systems, further contributing to deforestation. New technologies such as the plow, horse collars, and felling axes enhanced people's ability to cultivate difficult terrain, such as the moist, heavy soils of densely wooded areas.

The scale of deforestation in medieval Europe is hard to overestimate. Approximately 80 percent of temperate Western and Central Europe was covered in forests and swamps in 500 CE. Eight hundred years later, slightly more than half that area had been converted to agricultural fields, urban centers, grasslands, or other anthromes. One famous historical text gives us a sense of the extent of this shift. In 1086, William the Conqueror commissioned a systematic survey of much of England and parts of Wales. The king's goal was to determine what taxes were owed to the Crown, the types and extent of resources available, and where the people's loyalty lay. The assessors' reckoning of holdings and their values were recorded in the Domesday Book. This document is unique for the information it yields about medieval English life. According to its assessments, by 1086 only 15 percent of the English landscape was forest.

The construction of Notre-Dame began a few decades later, in 1163, and was substantially completed in 1345. Its timbers were cut from massive European beech trees three to four centuries old, which could still be found in refugia in France despite widespread deforestation. Just two years after the final stone was laid at Notre-Dame, the spread of bubonic plague across Europe between 1347 and 1353 wiped out at least one-third of the population, some twenty million people. The population pressures that had caused an agrarian crisis diminished almost overnight. Upwards of 25 percent of pastureland and farm fields were abandoned and quickly reverted to forest. Little changed over the next century, as war and famine devastated the European countryside.

By the end of the fifteenth century, however, Europe rebounded and began a long period of economic growth, commercial expansion, and global enterprise. Columbus's landing in the

Americas in 1492 marked the beginning of the Age of Discovery. The colonization of the New World, Africa, and Asia fundamentally changed social, economic, and political systems, pushing Europe toward a more entrepreneurial, commercial, and profit-driven economy. The intensive cultivation of crops such as sugar, cotton, and tobacco on colonized lands, along with global consumption patterns, gave rise to merchant empires whose leaders exerted almost regal powers. The farming of these crops depended almost exclusively on enslaved laborers, driving record profits for the owners of New World plantations, mines, and many other commercial enterprises. Global trafficking of humans became a commercial business unto itself. Millions of African people were captured, enslaved, and sold as New World laborers. The legacy of the global slave trade continues to shape social, economic, and cultural dynamics around the world today.

Merchant shipping activity expanded from an estimated 600,000 tons in 1600 to nearly 3.4 million tons by the late eighteenth century. A monetary economy replaced the barter system and facilitated global interchange, credit and banking systems, and an increasingly sophisticated financial infrastructure. Social networks connecting peasant farmers to local artisans, which had come to dominate exchange during the Middle Ages, dissolved. In their place arose more flexible, impersonal, and increasingly transnational and corporate exchange systems. Europe's population skyrocketed from 82 million in 1500 to 140 million in 1750. This increase created a high demand for commercial products, and the influx of gold and silver bullion from across the oceans led to a greater demand for high-value luxury goods.

All of these factors led to a renewed pressure on European forests. By 1700 agricultural fields covered 100 million hectares (about 250 million acres) across western, central, and northern Europe, mostly created by clearing forests. Shipbuilding to supply naval armadas, vessels of exploration, and transoceanic merchant ships created a demand for prime oak beams for masts and hulls. Nobility and royalty reserved some tracts of prime forest for game and sport, but these reserves only increased pressure on other remaining forest areas. With the onset of the Little Ice Age from about the early fourteenth century, colder winters brought the expansion of glaciers and snowfields, frozen rivers, and reduced groundwater. Broadleafed deciduous trees (e.g., beech, oak, elm, and ash) could not thrive in these conditions and were replaced by conifers in parts of Europe. The supply of timber diminished and prices rose, especially around urban centers in western Europe.

Millions of acres' worth of timber had been fed to domestic hearths and industrial furnaces before the shift to coal in the eighteenth century. The Industrial Revolution only accelerated European deforestation. By the nineteenth century, the world had become an interconnected marketplace. Land around the globe was cleared for cash crops, populations skyrocketed, and urban centers boomed. Industrial activities and mining increased demand for raw materials and energy. At the dawn of the nineteenth century, only 5 to 10 percent of Europe remained forested.

REPAIRING NOTRE-DAME

I had to wait nearly a week to make my pilgrimage to Notre-Dame. Slightly disoriented, I walked through a maze of majestic administration buildings, cafés, temporary fencing, and armed policemen to the southeast end of the Île de la Cité.

Twenty years earlier, the island had been a tourist epicenter. On this cold October morning, in the midst of a pandemic, the cafés were largely empty, and the tourist shops were deserted. Most of the people

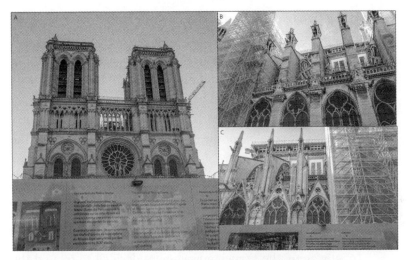

View of the west facade of the cathedral of Notre-Dame de Paris under restoration in October 2021. After the devastating 2019 fire, the once-brilliant stained-glass windows are charred black, the iconic spire is missing, electrical cords snake up the building facades, and a fence topped with barbed wire bars entry to the cultural center of Paris.

on the streets were lined up in front of administrative buildings with paperwork folded in their hands or tucked into jacket pockets.

When I reached the cathedral, I was immediately struck by the charred, blackened look of the stained-glass windows and the remains of the roof. Although I had seen plenty of pictures of the fire damage, I was unprepared for the profound sadness and emptiness I felt. I knew that Notre-Dame's iconic spire had been lost, but I was still shocked to see how its absence changed the entire character of the cathedral. The once-brilliant stained glass was dark and foreboding, as if Notre-Dame's light had burned out.

A construction fence topped with barbed wire enclosed the cathedral. Thick electrical wires snaked their way up the sides of the cathedral, and a massive crane loomed, unmoving, on the south side. Scaffolding towered above, and electric drills buzzed constantly.

Interpretive signage gave tourists some information about the construction history and the meaning of Notre-Dame's religious iconography. More abundant, however, were the colorful paintings and drawings by schoolchildren, reminding viewers of the Notre-Dame that once was and hopefully will be again.

The signage provided details of the 2019 fire and the efforts to save as much as possible of the original structure and repair the damage. Several sections of roof had been constructed from new timbers and bolted into place. I could tell immediately where modern construction materials had been incorporated into the framework, but I could not tell whether these would be visible once the restoration was finished.

In July 2020, the French government decided to rebuild the roof and spire as replicas of the originals, despite some proposals to construct a contemporary glass spire, a rooftop garden, and other modern touches. Approximately two thousand oaks, more than two centuries old, will be felled from the Forêt de Bercé in northwest France, where oak and beech trees have been sustainably managed. National forests like this, along with small private forest holdings, represent some of the last vestiges of long-lived trees in Europe.

The lesson of that moment was a sobering one. Climate change, anthropogenic impacts, and environmental crises threaten not only the beauty of rugged, wild places and the brilliance of plant and animal biodiversity around the globe but also monuments and cultural centers. These places help us remember who we are, where we come from, and how we got here. My trips to Paris and Notre-Dame as a high schooler and young adult left an indelible impression on me. I can't imagine a Paris without Notre-Dame, just as I can't fathom a world without the Brazilian rainforest. To rally support in combating our environmental crisis, we must help everyone understand what is at stake.

FORESTS TODAY AND TOMORROW

Since the end of the last Ice Age, humans have reduced the global number of trees by an estimated 46 percent. Thirty percent of the world's land area is currently covered by forests, but they are rapidly disappearing. According to the World Bank, between 1990 and 2016 we lost an area of woodlands larger than South Africa—1.3 million square kilometers (502,000 square miles). Seventeen percent of the Amazonian rainforest has been felled or burned in the past fifty years, and losses are accelerating. Fifteen billion trees are cut down every year.

Forests are more than just places to have a picnic or hang a hammock; we need trees for a variety of reasons. Perhaps most importantly, trees are the natural air filters of our planet. They absorb carbon dioxide and other heat-trapping greenhouse gases emitted through activities like driving cars, operating factories, running air conditioners, and even breathing. Through their own metabolic processes, they emit oxygen. In a single year, an acre of mature trees can absorb the quantity of carbon dioxide produced by driving your (gas-powered) car twenty-six thousand miles and emit enough oxygen to support eighteen people.

Trees also make a neighborhood a cooler, less expensive, and friendlier place because of the shade they create and the water vapor they release. In the last fifty years, average temperatures in Los Angeles have risen by about 6°F as tree cover has been removed to make way for roads and buildings, which retain more heat. Breaking up concrete jungles with trees can cool a city by up to 10°F. Placed strategically around a home in an American suburb, according to the United States Forest Service, trees can cut a summer air-conditioning bill by up to 50 percent, reducing local energy demand as well as carbon dioxide emissions and other pollution. Sociological

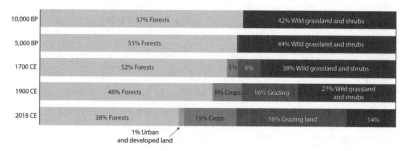

Simplified graph of global landcover changes from the dawn of domestication and agriculture until the twenty-first century. Ten thousand years ago, 71 percent of Earth's land surface was covered by forests, grasslands, and scrubland, for a total of 10.6 billion hectares (26.5 billion acres). Nearly half (46 percent) of this land is now used for agriculture.

studies also suggest that access to green space and trees can help people recover from mental fatigue, build social bonds, and reduce neighborhood crime, violence, and aggression. We need to view trees as more than just inputs for human industry and recognize that the beauty they bring to the world and our lives is critical for our well-being.

Deforestation around the world has a long, complicated history, which we must understand if we are to reverse the trend and restore forest landscapes. Archaeology, palynology, written texts, and other historical data can tell us where forests once stood, their species compositions, when and under what conditions they thrived, and the ways they changed over time. Interdisciplinary efforts that make use of deep historical data can help us combat climate change quickly and effectively, in ways that benefit our lives much more than simply reducing global warming. Restoring forests and other wild places contributes to a healthy planet and healthy human communities.

In November 2021, leaders from more than one hundred countries around the world, including Brazil, China, Russia, and the United States, committed to end deforestation by 2030. This is certainly an important step: there is no path to net-zero carbon emissions without curbing tropical deforestation. Governments, corporations, scientists, and citizens must find ways to follow through with this promise. A similar declaration was made in 2014 with a zero-deforestation target date of 2030, but forest loss was 41 percent higher in the years directly after the announcement than before it. But we can still succeed. Global deforestation rates have declined every decade since their peak in the 1980s. Individuals can make a difference. Plant a tree. Go paperless at home and the office. Buy recycled products and recycle the products you have. Buy only wood products certified by the Forest Stewardship Council, a nongovernmental organization that promotes responsible and sustainable management of the world's forests. Support companies that commit to reducing deforestation. Tell a friend and raise awareness in your community. Most importantly, support leaders and organizations that recognize the problem and commit to doing something about it.

○

Chapter 5

THE DISAPPEARING
BIRDS OF HAWAI'I

W hen you live in sunny San Diego, taking a vacation can sometimes feel strange. We have beaches, some of the most beautiful in the world. We have sunshine on average 266 days a year. Even the worst winter days are pleasant compared to those in most of the United States. We also have a lot to do. Drive an hour east, and you are in the mountains. Drive thirty minutes south, and you are in Mexico. Eight miles northeast of downtown is Mission Trails Regional Park, one of the largest urban parks in the United States. We have great food, a range of live entertainment, and some of the best museums and cultural institutions in the country.

The number one attraction for visitors is probably the San Diego Zoo. Located in the heart of the city, it is home to more than 3,500 animals and hundreds of thousands of plants. In a single day, you can see leopards, elephants, koalas, African penguins, Galápagos tortoises, pygmy hippos, hamadryas baboons, polar bears, Tasmanian devils, and more plant species than I can name. For San Diego residents, a family zoo pass is one of the best deals in town. My wife and I take our son to the zoo at least a couple of times a month. We always visit the reptile house (snakes are his favorite), but the options are endless—kangaroos, bears, penguins, insects. We never know what will capture his attention. I, however, always root for the birds. Not the raptors like eagles, hawks, falcons, and owls that tend to fascinate a six-year-old, but the songbirds and the tropical birds, in all their dazzling colors, shapes, and varieties.

I am not an ornithologist and would not even call myself a birder. But, as years pass, my aching knees and lower back have taught me to appreciate the opportunity to sit in the various bird enclosures and simply watch. The William E. Cole Family Hummingbird Habitat is my favorite. This enclosure is complete with cascading waterfalls, an open-water pool, colorful orchids and anthuriums, palms, and a variety of rare plants, along with several benches where visitors can observe the whirling of wings and the cacophony of calls from seventeen different bird species. Tanagers and hummingbirds swoop from one feeder to another. Other residents seem unaffected by all the activity, like the wattled jacana, with their chestnut back and wing feathers, black bodies, and brilliant red head shields, majestically balanced on one leg in calm pools while they probe the water with yellow bills. This juxtaposition of activity and serenity is hard to beat. Every visit brings me a renewed appreciation of the vast diversity of species, evolutionary adaptations, and behaviors on display.

Despite all the local activities and nearly perfect weather, in the winter of 2021 my family decided to take a trip to Hawai'i. Again, it might seem strange to leave San Diego for an island and beach vacation, but after staying home for many months during the COVID-19 pandemic, a change of scenery was in order. We settled on six days in Kaua'i, the Garden Island, the lushest, calmest, and (perhaps) prettiest of all the Hawaiian islands, with stunning natural features and abundant hiking trails. For me, an added bonus was the possibility of seeing some rare and unique birds. Even though it is one of the smaller islands, Kaua'i is second only to the Big Island in avian endemics, or species that occur naturally only in that place. There are six bird species found only on Kaua'i. I was excited about the chance to see these birds in their native habitat.

As often happens, reality did not match my expectations. On Kaua'i, the most ubiquitous bird—present in every parking lot, beach, and forest on the island—is the chicken. According to local lore, the Kaua'i chickens are descended from birds that escaped into island jungles when their coops were destroyed during Hurricane Iwa in 1982 and Hurricane Iniki in 1992. Feral chickens are found on other Hawaiian Islands, but not in the large numbers that they are found on Kaua'i. These chickens have become the unofficial mascot of the island. To be fair, they are not the drab egg-laying machines you would expect to find on a mainland farm. They are quite colorful and active, and my son, at least, was delighted by their omnipresence. We came home with a souvenir refrigerator magnet for our collection, adorned with a regal-looking rooster. I went to Hawai'i to observe the richness of native birdlife, and I came home with a chicken magnet.

Until recently, the Hawaiian Islands truly were a wonderland of exotic and colorful birds, most of which have now been lost forever. Before the arrival of humans, 142 distinct bird species found nowhere else on the planet inhabited every ecological niche in the islands. Over millions of years, they adapted to changing environments. They diverged, evolved into separate species, and developed new feeding habits, beak shapes, and behaviors; and they took full advantage of the scarcity of native predators. We know of these species only because their bones and fossilized remains have been found in lava tubes, paleontological deposits, and archaeological sites. From flightless ibises to bird-hunting owls and a spectacular variety of honeycreeper, the diversity was astonishing. Today,

101

ninety-five of those magnificent bird species are extinct; of those that remain, thirty-three of forty-seven (roughly 70 percent) are listed as endangered species. One-third of the endangered bird species in the United States are Hawaiian. The prospects for many of the surviving native Hawaiian birds are not encouraging. Introduced diseases, habitat loss, and the introduction of nonnative predators such as rats, snakes, cats, and mongooses threaten their survival in the wild.

The real trouble for Hawai'i's birds started with humans. The first wave of extinction began with the arrival of Polynesian maritime voyagers some 1,600 years ago. Flightless megaducks, geese, and rails were easy prey for Polynesian hunters, and their eggs and young were targets for the rats, dogs, and pigs Polynesians brought with them. High-ranking Polynesian chiefs (*ali'i*) decked themselves in lavish cloaks (*'ahu 'ula*) and helmets (*mahiole*) constructed from the colorful feathers of local birds, which were believed to provide physical and spiritual protection for their wearers. Each cloak and helmet required hundreds of thousands of feathers to manufacture. Although birds were not killed for this endeavor, but rather captured in order to pluck their feathers and then released back into the wild, the practice may have stressed local avian populations. The clearance of native forests by Polynesian farmers also stressed bird communities and likely contributed to extinctions.

Captain James Cook's arrival at Waimea Bay on Kaua'i in 1778 jump-started a second, continuing wave of avian extinctions. The establishment of European and American farming and ranching enterprises and urban development resulted in widespread habitat destruction and brought a new selection of introduced predators. These included the black rat (*Rattus rattus*), which, unlike the Polynesian rat (*R. exulans*), can easily climb trees and raid bird nests. Perhaps the most devastating introduction was the mosquito, which led to deadly outbreaks of avian pox beginning in the mid- to late 1800s and avian malaria starting sometime in the early twentieth century. Today, only twenty-one native songbird species remain on the Hawaiian Islands; eleven of these are listed as endangered, and most are restricted to high mountain regions where mosquitoes cannot survive. And as the world warms from anthropogenic climate change, mosquitoes are moving into these high-altitude refugia, spelling potential doom for Hawai'i's endangered birds.

The challenges facing Hawai'i's avifauna are a part of a biodiversity crisis on a scale the Earth has not seen for over sixty-five million years. We do not know for certain how many unique plant and animal species are lost every day, because many species are unidentified and others, including entire regions and groups of plants and animals, are understudied. But computer models suggest that somewhere around 150 species are driven to extinction every single day.

THE SIXTH MASS EXTINCTION

From the perspective of geologists and climate scientists, the history of Earth's flora and fauna over the past 450 million years can be seen as a play with five acts. These consist of five mass-extinction events—episodes of sudden, dramatic change in the Earth's climate and environment, when at least half the planet's macroscopic plants and animals vanished. Geologists have used these events to demarcate geologic epochs or time markers in the Earth's history. Most of them took place so long ago and under such mysterious conditions that they are, for the most part, poorly understood by scientists.

The largest and most devastating mass extinction in Earth's history was the Permian-Triassic event 252 million years ago, which resulted in the extinction of over 95 percent of Earth's species, including 70 percent of land-dwelling vertebrates. So much life was wiped off the face of the planet that it has been colloquially dubbed the Great Dying. This event almost resulted in the end of life on our planet. Understanding how and why it occurred is difficult. The Permian-Triassic extinctions occurred in pulses. The earliest was the result of gradual environmental change, and the later pulses were likely triggered by an asteroid or comet impact, the Siberian Traps (massive volcanic eruptions that lasted roughly two million years), sea floor methane release, or some combination of these phenomena.

The most recent mass extinction event is the one with which you are probably most familiar—the Cretaceous-Paleogene or K-T boundary event. (The name *K-T* is derived from the German word *Kreide*, meaning "chalk," referring to the chalky sediment of the Cretaceous Period, and the word *tertiary*, traditionally used to describe the interval spanning the Paleogene and Neogene periods.) Most geologists attribute it to a massive meteorite or comet impact 65.5 million years ago near Chicxulub, Mexico, although others implicate changes in atmospheric carbon dioxide or tectonic plate movements. The K-T event resulted in the extinction of about 76 percent of terrestrial species within just a few millennia, including nonavian dinosaurs, early mammals, and amphibians, birds, reptiles, and insects. Also wiped out were marine reptiles like mosasaurs and plesiosaurs and one of the most diverse families of animals on the planet, ammonites.

Of course, extinction is part of life on Earth. For billions of years, new species of plants and animals have evolved on our planet, outcompeting and replacing other species, resulting in the rich biodiversity, past and present, that we know today. About 98 percent of all the plant and animal species that ever existed on our planet are now extinct. Generally, when a species goes extinct, a new one fills the empty ecological niche, and life marches on. When everything is operating normally, Earth's extinction rate is comparatively slow: between 0.1 to 1 species per 10,000 species over one hundred years. Known as the background extinction rate, this is part of our planet's functioning. Life on Earth steadily changes as living things adapt to gradually changing conditions.

Mass extinctions, by contrast, are disruptive and critical turning points in biotic evolution. They wipe out fit and unfit species alike and result in dramatic declines in biodiversity. Recovery from such events takes millions of years and results in the transformation of floral and faunal communities. It took more than ten million years for mammals to evolve into the diverse array of species that rivaled the diversity of the dinosaurs after the K-T boundary extinction event. During mass extinctions, new species cannot evolve fast enough to fill the ecological role performed by the extinct species. Their absence can result in cascading effects throughout ecological and planetary systems.

A growing consensus among the scientific community is that we are living in the midst of another mass extinction. Since it is occurring around us, it is impossible to predict the results or know how it will compare to the previous "Big Five." But according to current calculations, the rate of modern extinctions is one hundred to one thousand times higher than background levels. At this rate, the sixth extinction could result in a 50 percent loss of the remaining plant and animal life on Earth. This loss in biodiversity would be catastrophic. It could trigger the collapse of entire ecosystems. For humans it could spell the loss of critical food economies, the disappearance of medicinal and other resources, and the demise of important cultural landscapes and seascapes. Many species are teetering on the brink: extinctions threaten one-third of amphibian species, nearly one-third of corals, one-quarter of all mammals, and one-eighth of all birds.

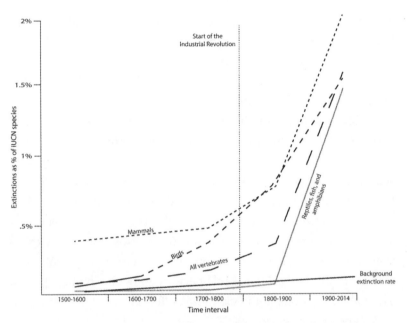

The percentage of species determined to be extinct or extinct in the wild by the International Union for the Conservation of Nature (IUCN) from 1500 BCE to 2014 CE. Note the dramatic increase in extinction rates compared to expected background rates. The authors from whose work this graphic was derived state that they used a conservative approach, and it likely *underestimates* the severity of the modern extinction crisis.

Extinctions of plants and animals can have far-reaching implications for the health and functioning of our world. Coral reefs, for example, protect coastlines from the damaging effects of wave action and storms. They are also the source of nitrogen and other essential nutrients for marine food webs and of food and new medicines for human communities. Their loss would be devastating to life on the planet.

The undeniable cause of our current mass extinction crisis is us. Humans have already transformed more than 70 percent of Earth's land surface and have put to use about 75 percent of the world's freshwater. Industrialized agriculture often results in soil degradation, deforestation, pollution, runoff, and biodiversity declines. Plowing disrupts soil ecologies. Planting vast fields with a single crop (mono-cropping) drives out species from their natural habitats and creates conflicts between humans and wildlife.

Some species are known as *keystone* species because of their critical importance to specific ecosystems. These include wolves, which help keep the population of grazing animals in check, and beavers, whose dams help regulate water flow in river ecosystems. When humans target these species, as pests or as resources, their removal can cause cascading effects across entire habitats.

If removing species can be destructive, so can introducing new species in the wrong place. Humans have been responsible for transporting invasive species across the globe—sometimes delib-erately, sometimes accidentally. Introduced species compete with native flora and fauna for resources and often reduce local biodi-versity. When native species are ill-equipped to compete with the new arrivals, the result may be extinctions of rare native plants and animals. Finally, human contributions to greenhouse warm-ing and climate change have triggered environmental changes at a scale and rate to which many species are unable to adapt. The International Union for the Conservation of Nature (IUCN) esti-mates that because of anthropogenic climate change, nearly eleven thousand species on their Red List of Threatened Species face increased likelihood of extinction.

Bad as this news is, I see a silver lining. The mass extinctions of the deep geological past occurred long before our earliest human ancestors appeared about seven million years ago. The Big Five mass extinctions were all caused by extreme temperature changes, dra-matic fluctuations in sea levels, or catastrophic one-off events like colossal volcanic eruptions or extraterrestrial objects hitting the Earth. There is little we could do to prevent or mitigate the effects of an event of this kind. But since the current mass extinction is a crisis humans have created, that also means we can fix it. We can find ways to live more sustainably and reduce the loss of biodiversity around the world. We can study the causes and effects of human-induced plant and animal extinction and work to curb them. We can identify our most harmful activities and find alternatives or ways to reduce the negative effects. Doing so is important not just for the plants and animals being driven to the brink of extinction (and beyond): it is important for the quantity and quality for our time on Earth.

Since the sixth extinction is often strongly linked to recent or modern human activities beginning with the Industrial Revolution, the study of deep history may seem to offer little information relevant to the current crisis. However, humans have driven extinctions, at a slower rate, for millennia. Many of these have shaped the character and composition of modern land- and seascapes. Understanding the roots of the sixth extinction, using archaeological and paleoecological evidence, is critical for stemming the tide of plant and animal extinctions and preserving biodiversity. Lessons from history allow us to find ways to help and to evaluate whether our efforts are effective.

THE MIAMI BLUE BUTTERFLY: CHARTING CONSERVATION FROM HISTORY

In Florida, two biological scientists used innovative field research and historical detective work to help save the Miami blue butterfly, a stunningly beautiful insect. Both male and female Miami blues have front wings measuring only about half an inch in length each. The metallic blue backsides of their wings are eye-catching when opened up or in full flight. By contrast, the gray undersides of the wings allow them to blend into their surroundings. Both sexes display black coloring along their hindwing margins and black spotting on their wing undersides. These creatures are lookers. Unfortunately, they are also extremely rare.

Not long ago, Miami blues could be found inhabiting tropical hardwood canopy forests, tropical broadleaf forests, and beachside scrub environments along more than half the coast of southern Florida (on both the Gulf of Mexico and the Atlantic Ocean) and the Florida Keys. The species was a familiar sight to adventurous hikers, beachgoers, and nature lovers until populations began to plummet in the 1980s. By the end of the decade, the butterflies could be found in only a few locations along the Florida Keys. By 1992, scientists knew of only one remaining colony, on Key Biscayne. That year Hurricane Andrew, one of the most destructive hurricanes ever to hit the United States, made landfall in Florida, with sustained winds of more than 156 miles per hour. Hurricane Andrew wiped out the Miami blue butterfly colony on Key Biscayne, leaving scientists and butterfly enthusiasts desperately searching for any surviving remnants.

The collapse of Miami blue butterfly populations was the result of multiple factors, the most powerful of which can be directly attributed to human activities. Along mainland coastal Florida, Miami blues were eradicated from their historical range as urban and suburban developments felled native forests and scrublands and transformed Florida into the densely occupied vacation and retirement wonderland we know today. At the same time, Florida's decades-long war on mosquitoes spewed toxic chemicals into the environment. While these efforts successfully reduced the target pests, along with the yellow fever and dengue fever they spread, Florida's butterflies and other insects were collateral damage.

After years of searching in the aftermath of Hurricane Andrew, a small population of Miami blues, perhaps as few as fifty adults,

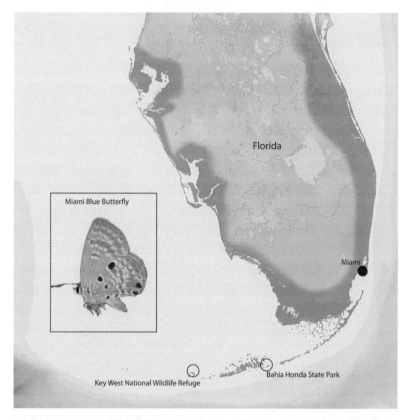

Map depicting the historical range (shaded area) of Miami blue butterfly populations throughout coastal southern Florida and the Florida Keys prior to their catastrophic decline. Two refugia habitats in the southwestern Florida Keys—Bahia Honda State Park and Key West National Wildlife Refuge—are marked with open circles. (Image via Wikimedia Commons.)

was found on Bahia Honda Key in 1999. This discovery sparked a grassroots effort to save the species. After several petitions and years of legal haggling, the Miami blue butterfly was granted federal endangered-species status on an emergency basis in January 2003. Since then, considerable money has been raised to support its recovery, and a captive breeding program was established. Despite these efforts, the Bahia Honda colony disappeared in 2010, likely from the combined effects of drought, cold temperatures, and predation by nonnative green iguanas.

All was not lost, however, as two small colonies had been discovered in 2006 in the Key West National Wildlife Refuge and the Great White Heron National Wildlife Refuge. These populations are the focus of monitoring and recovery efforts today. They remain at high risk of extinction because of stressors like droughts, temperature fluctuations, storms, diseases, and parasitic outbreaks.

The blue butterfly was almost lost forever in part because conservation managers initially lacked information about the ecology and feeding habits of the species. Using a combination of museum data, historical distribution data, and modern field investigations, two entomologists from the University of California at Davis, Scott Carroll and Jenella Loye, set out to understand the reproductive ecology of the Miami blue butterfly and its decline by taking a long-term perspective.

One of their key findings came from museum herbarium collections used to map the historical changes in the range of balloon vines, a type of climbing plant that is widely distributed across tropical and subtropical areas of the world. Carroll and Loye discovered that balloon vine was once found from northern peninsular Florida down through the Florida Keys, all the way to Key West. By integrating historical and modern data, they determined that the butterflies were commonly found in association with balloon vine. Carroll and Loye realized that multiple species of balloon vine were found in Florida, and resource managers had mistakenly believed all of them to be invasive. In fact, one species of balloon vine is actually a native, and its seeds are a critical food source for Miami blue butterflies. Efforts to eradicate balloon vine, along with declines related to activities like land clearance for housing construction, had unintentionally spurred the loss of Miami blue populations throughout southern Florida. The researchers' painstaking work to track the ecological relationships between Miami blue butterflies and Florida's plant communities demonstrated that this species of balloon vine was not an invasive weed but a plant essential to the butterflies' survival.

Jaret Daniels is a curator at the Florida Museum of Natural History, a professor in the Department of Entomology and Nematology at the University of Florida, and one of the foremost authorities on Florida insect life. He and his research group have spent many years working on the recovery and restoration of the Miami blue butterfly. To publicize and support their restoration efforts, they have created an interactive exhibit, in collaboration with the Florida Museum, to raise awareness about the plight of Miami blues. They even launched a brand of beer, Miami Blue Bock, brewed by the First Magnitude Brewing Company. Daniels has also developed a captive breeding program for Miami blues and is optimistic that they can be reintroduced to native habitats along the coast of mainland Florida. He views one of the challenges facing Miami blues as a common one for species that naturally have limited ranges and low population numbers: it is only when a species becomes exceptionally rare that the race to understand its ecology and life history and save it from extinction ramps up. That leaves researchers and conservation managers with precious little data to guide their recovery efforts. He regards the application of historical data to discover the critical links between the native balloon vine and the Miami blues as key in saving the species from extinction. This approach to restoration biology, while not yet common, is, Daniels believes, "the wave of the future." The fact that Miami blue butterflies survive and continue to contribute to our planet's rich biodiversity is a testament to its importance.

THE SLAUGHTER AND RECOVERY
OF SEALS IN THE PACIFIC

Archaeological records are critical for helping conservation scientists, policymakers, and the public support the recovery and restoration of species threatened with extinction. This work focuses on carving out space, resources, and opportunities for native flora and fauna to recover and repopulate, especially after large disturbances such as oil spills, agricultural activities, logging, and overhunting. A recent study of four hundred such cases concluded that "passive recovery" (simply allowing natural processes to play out after the removal of a cause of disturbance) should be the first option in trying to repair Earth's damaged ecosystems. The famous entomologist E. O. Wilson—often called the "father of biodiversity"—suggested that the best approach to conserving species and saving nature was to set aside half the planet as protected areas. The idea is that if we leave species and ecosystems alone and give them enough time and space, they can effectively recover along a natural trajectory. This strategy is often effective. Sometimes, however, archaeological and other deep historical datasets reveal surprising complications.

One of most dramatic success stories in wildlife recovery in the Americas comes from the Pacific Coast, where seal and sea lion populations were driven to near extinction by nineteenth-century commercial and recreational hunters. Fur traders initially came to the region to collect sea otter pelts (see chapter 6), but once otter populations were driven to near extinction levels, they turned their attention to seals and sea lions. Marine mammals were hunted not only for fur but also for their blubber, which was rendered and used as lamp oil. Among the seal and sea lion species targeted along the Californian and Mexican Pacific coasts were Guadalupe fur seals and northern elephant seals.

There are nine different fur seal species in the world. Eight belong to the genus *Arctocephalus* and are found primarily in the Southern Hemisphere. One, the northern fur seal, belongs to the genus *Callorhinus* and lives in the Northern Hemisphere. Guadalupe fur seals (*Arctocephalus townsendi*) are the only species of southern fur seal that lives north of the Equator, concentrated today on the remote Isla Guadalupe off northern Baja California, Mexico. Because it was not recognized as a distinct species until 1897, long after commercial sealing had devastated the population, little is known about its previous geographic range and natural history, but some estimates place the species' prehunting numbers at more than two hundred thousand individuals. These animals have a dense undercoat that became highly prized in the garment trade during the nineteenth century—so much so that sealers slaughtered every animal they encountered.

Because of this carnage, the species was twice thought to be extinct. No Guadalupe fur seals were sighted from 1895 until 1926, when sixty adults and pups were seen, or from 1928 until 1949, when a lone male was discovered. Fortunately, a group of seals must have evaded the sealers. In 1954, a small breeding group of Guadalupe fur seals was discovered in a cave on Isla Guadalupe, and in a little over a decade their numbers had grown to five hundred individuals. They came under federal protection and had further opportunity to recover with the passing of the United States Marine Mammal Protection Act of 1972 (established to protect marine mammal species and populations from catastrophic declines and to provide resources for scientists to study them and aid in their recovery) and after the Mexican government declared their home island a seal and sea lion sanctuary in 1975. By 1993, populations had grown to about 7,400 animals, and in 2003 they reached 12,176. Today, population estimates place their numbers around 40,000, and Guadalupe fur seals are listed as a threatened species, but not as endangered or under dire threat of extinction.

Unlike fur seals, northern elephant seals have skin with stiff, coarse hair that offered little to the commercial pelt trade. These enormous animals, however, have copious amounts of blubber, which eighteenth- and nineteenth-century hunters collected for oil, especially after commercial hunting devastated whale populations. First described as a species in 1866, northern elephant seals were thought to be extinct by the late 1870s, until a series of research expeditions between 1880 and 1884 found a remnant population on Isla Guadalupe. The species was again thought extinct until 1892, when nine animals were located on Isla Guadalupe. Despite being hunted to near extinction, isolation and a cessation of commercial hunting allowed northern elephant seal populations to rebound fairly quickly. Even before federal protections were enacted in the 1970s, northern elephant seals continued to rebound and expand their geographic range, establishing breeding communities on the Channel Islands off the coast of California by the 1950s and steadily moving up the Pacific coast into central California by the 1970s, the San Francisco Bay Area and Oregon by the 1980s, and Washington State and British Columbia by the early 2000s. Populations of northern elephant seals today are estimated at around two hundred thousand animals, comparable to their estimated prehunting numbers. This population growth, combined with an expanded biogeographic range, makes the northern elephant seal's recovery a major conservation success story. Today, you can view these gigantic creatures hauled out and sunbathing in large groups on numerous beaches along the US Pacific Coast.

Marine biologists have assumed that Guadalupe fur seals and northern elephant seals are repopulating the US and Mexican Pacific coast much as they occupied it before historical overhunting.

Archaeological evidence suggests, however, that this may not be the case. In a series of studies directed by the Smithsonian Institution archaeologist Torben Rick, at least sixty Native American archaeological sites on the California coast, from San Diego County to San Mateo County, have produced Guadalupe fur seal remains—more than 3,400 bones and teeth. While sea mammals were not a consistent or ubiquitous component of Native American diets, coastal Indigenous groups in California hunted them for thousands of years as a source of protein and raw materials. The oldest remains are at least eleven thousand years old, and the most recent date to after European contact. In contrast, just twenty-eight archaeological sites ranging from Baja California to southern British Columbia have produced northern elephant seal remains. Of the ninety-six northern elephant seal bones and teeth identified, the oldest date back to about seven thousand years ago and the most recent to after European contact. In a more recent study that employed cutting-edge chemical identifications of small sea mammal bone fragments from several Northern Channel Island archaeological sites, elephant seal remains also were identified in an archaeological deposit approximately twelve thousand years old.

By the measures of modern ecological studies, efforts to restore the populations of Guadalupe fur seals and elephant seals have been spectacularly successful. In little more than a century, two species that scientists believed extinct or on the verge of extinction now number in the tens or hundreds of thousands. Guadalupe fur seal populations stand at 20 percent of their estimated pre-fur trade numbers, with annual increases of around 10 percent, and elephant seal populations have reached or exceeded prehunting levels. Three US federal agencies were involved with the successful recovery efforts: the National Marine Fisheries Service (often referred to as NOAA Fisheries), US Fish and Wildlife Services, and

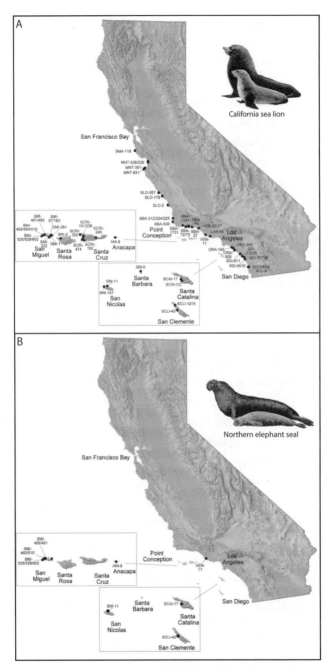

Maps showing the locations of prehistoric archaeological sites in southern and central California that contain remains of (a) Guadalupe fur seals and (b) elephant seals. These data suggest a major shift in the geographic ranges of these animals from ancient to recent times, likely the result of overhunting during the commercial fur trade. (Images courtesy Torben Rick.)

the Marine Mammal Commission. But a longer view of these populations along the Pacific coast of North America suggests that something is amiss. Beaches teem with elephant seals where they were largely absent in the past. Guadalupe fur seals, relatively common in California waters for thousands of years, are rare visitors today, perhaps because of elephant seal encroachment.

These data, based on millennia-long patterns found in the archaeological record, raise important questions. Why have the biogeography and relative abundances of these two species changed? Are the changes related to changing ocean conditions, food availability, or something else altogether? Will Guadalupe fur seal numbers continue to increase, and if so, will they extend their range into more northerly waters? How are California marine food webs different now, with abundant elephant seals, than they were in the past, with abundant Guadalupe seals? Are these modern shifts an indicator of a larger problem or dysfunction in the oceans? Was one species simply able to recover more quickly than the other?

One of the most experienced and knowledgeable marine mammal biologists in the world, Robert "Bob" DeLong, is a former head of the Alaska Fisheries Science Center's California Current Ecosystems Program and recently retired from NOAA. Bob is a strong advocate for using archaeological and historical data to investigate the state of sea mammal communities in the Pacific. I first met him while conducting an archaeological survey on San Miguel Island, in the California Channel Islands, as part of my graduate studies at the University of Oregon. He has over forty years of experience working on San Miguel.

My PhD adviser, crew, and I made the hike of five or so miles to the far western end of the island in the early morning to meet up with Bob and explore the southwestern coast, where access is strictly limited. He built and maintained a small research bunkhouse that overlooks sandy, windswept Point Bennett, one of the largest seal and sea lion rookeries in the world. As we interacted over the years and our friendship grew, I was occasionally invited to stay there when conducting research on the west side of the island. San Miguel is a difficult place to live comfortably, but through effort and ingenuity Bob made it feel like home.

Bob is a repository of knowledge, stories, and lore about San Miguel and its natural and cultural history. He can tell you how to administer an enema to a two-ton sea mammal or about the time a bull elephant seal bit down on his arm and flung him around like a rag doll as he distracted the animal from charging a colleague. He can also recall what Point Bennett was like when only a handful of northern elephant seals, harbor seals, and sea lions could be found there, as opposed to the more than one hundred thousand that occupy the area today. Bob was working on the island prior to the formation of Channel Islands National Park in 1980, shortly after ranching operations ended. His wealth of knowledge and island experience may be unmatched.

Bob has spent a lifetime studying these animals, from the diseases that affect them to their reproductive strategies and mating behaviors to their foraging activities. When I asked him why Guadalupe fur seals have been slow to recover along the Channel Islands, whereas elephant seals have made a rapid recovery, he was as curious as I was. He explained that Guadalupe fur and elephant seals both regularly feed offshore in oceanic waters, but Guadalupes do not dive below 200 meters, whereas elephant seals forage farther offshore and to depths of about 450 meters. Guadalupes feed primarily on squid and small pelagic fish such as mackerel, sardines, and lanternfish; elephant seals feed in deeper waters, preying on squid along with deepwater fish and even some sharks and rays. This suggests that the two species are not in direct competition with one another. Bob believes their divergent numbers may be due more to chance than to anything else. Elephant seals had a head start in expanding their numbers and range by gaining a foothold on largely empty beaches and in open ecological niches. But we do not know for certain.

The spectacular recovery of seals, sea lions, and other sea mammals from the brink of extinction seems like a momentous success story. These animals survived a campaign of deliberate mass slaughter and can now be found all along the US Pacific coast. Yet the aftermath of overhunting still threatens their continued survival. Because hunting reduced their populations to such tiny numbers, they have lost much of their genetic diversity, potentially making them more susceptible to disease, environmental shifts, and other stressors. So far, little research has investigated the scale of this reduction in genetic diversity and its potential consequences. Bob believes that new viral or bacterial infections will emerge among sea mammal communities in coming years, especially given the rapid changes in oceanographic conditions precipitated by anthropogenic climate change. Low genetic diversity could impair the species' ability to survive.

Genetic studies of sea mammal bones from archaeological and paleontological sites, museum collections, and modern animals can help assess the extent of reduced genetic diversity and what this might mean for the species' adaptive fitness. Chemical analyses of bones can also help reconstruct the feeding behaviors and preferred prey of ancient marine mammals and detect potential changes or shifts over time. Marine shells recovered from archaeological sites contain information on local oceanic conditions in the deep past, which can shed light on how sea mammals might fare under future oceanographic conditions. Scientists will never be able to bring back plants and animals that have gone extinct, but they can help build more effective management plans to protect plants and animals that still remain.

PRESERVING BIODIVERSITY

The first step in preserving and protecting plant and animal species around the world is understanding how important our planet's biodiversity is for our own lives and recognizing how little we know. The extinction of a species is like removing a brick from a massive tower. Every brick is part of the structure of the tower, but losing a few does not necessarily result in the tower's collapse. However, as more and more bricks are lost, more and more pressure is placed on the others, until eventually the structure fails and the tower collapses. One of the major challenges to preserving biodiversity is that there are an estimated one hundred million plant and animal species on Earth. Even with thousands identified every year, fewer than 2 percent of these have been documented, described, and cataloged by scientists. We don't know how many and which bricks are being lost. Without a detailed understanding of the vast diversity of life on Earth, our best hope is to preserve and protect biodiverse-rich environments and ecosystems, helping to maintain the species diversity we know along with what we have yet to identify.

We lose biodiversity by encroaching on the places where diverse plants and animals thrive, such as mangrove, deciduous, and tropical forests; estuaries and wetlands; coral reefs; and kelp forests. This is bad for biodiversity and can also be very bad for us. For at least a century, an average of two new viruses have passed from animal hosts to humans every year. As I write this in September 2022, these include COVID-19 and mpox (monkeypox), whose transmission to humans has been linked to human-driven environmental change such as deforestation and the wildlife trade that bring vectors, pathogens, and hosts into greater contact with humanity and can lead to the emergence of zoonotic diseases (diseases transmitted from animals to humans).

The economic cost of protecting biodiversity is orders of magnitude less than the cost of global pandemics such as COVID-19. By preserving biodiversity and species-rich habitats, humans around the world can also enjoy better health, cleaner drinking water, a supply of diverse foods and medicines, and all the positive mental health effects that come from re-creating and enjoying natural places.

Preserving biodiversity is also an excellent tool for combating the climate crisis and the production of anthropogenic greenhouse gases. Forests on land and under the sea (such as kelp, seagrass, and mangrove) absorb 60 percent of human-produced carbon dioxide emissions, totaling 5.6 gigatons of carbon absorbed by plants per year. The relationship between biodiversity and climate creates a global feedback loop. As we degrade diverse ecosystems, the climate changes, and the loss of biodiversity accelerates. As we protect ecosystems and slow or reverse the loss of biodiversity, we slow human-induced climate change. If we make the right decisions, nature-based solutions could account for more than one-third of the emission reductions necessary for nations to meet the goals of the Paris Agreement by 2030.

Preserving biodiversity is not only an important part of environmental sustainability but also an urgent social justice issue. The loss of our world's biodiversity disproportionally affects poor, marginalized, and Indigenous communities, as it is these groups who rely most heavily on the local environment for their food and livelihoods. All human communities rely on clean air and water and fertile soils. When access to these resources is disrupted, vulnerable communities have few alternative ways to secure essential human needs. Most immediately, the loss of biodiversity can cause food insecurity for local communities and deepen social inequalities. Poor communities around the globe that already struggle to access freshwater and food are positioned to suffer the most as Earth's biodiversity erodes.

This is especially the case for the Indigenous peoples who occupy a quarter of the Earth's land surface. Archaeology, oral histories, and other deep historical data demonstrate that Indigenous communities and their ancestors have influenced and helped to protect, manage, and shape ecosystems, biodiversity, and biodiverse-rich environments for millennia. Their knowledge and traditional practices often are closely tied to local biodiversity. Eroding this biodiversity not only harms the flora, fauna, and resources that have underpinned their economic systems but also disrupts cultural heritage and knowledge systems. In the Amazon rainforest, for example, Indigenous communities such as the Yanomami and Kayapó in Brazil, the Ashaninka in Peru, and many other groups face significant threats to their traditional lifeways and livelihoods due to deforestation, illegal logging, mining, and land encroachments. Indigenous Pacific Islander communities contend with rising sea levels, ocean acidification, and coral beaching that severely impacts marine biodiversity, fish populations, and traditional fishing practices. The list goes on and on. Indigenous communities stand to lose their homes, lifeways, and centuries of cultural history in the current mass extinction. Those communities that will suffer the heaviest losses are those least responsible for the erosion of biodiversity, anthropogenic climate change, and environmental impacts.

●

Chapter 6

THREE-MILLION-DOLLAR TUNA IN TOKYO

T he famed Tsukiji Market along Tokyo's waterfronts, the largest seafood market in the world, has become a destination for foodies and tourists. Thousands wander through its maze of stalls and restaurants to gawk at the variety of sea creatures (many flown or shipped in from sources around the world) sold cooked, fresh, or prepared to order. The Tsukiji Market has been a Japanese institution for centuries. It offers a sense of the world's rich diversity of sea life, from common seaweeds to extravagantly expensive varieties of caviar. Tsukiji is the literal and symbolic epicenter of commercial fishing and the place where global market prices are set for seafood delicacies.

At a predawn auction at the Tsukiji Market in January 2019, a 612-pound (278-kilogram) Pacific bluefin tuna sold for a record-setting 333.6 million yen (about US $3 million). It was purchased by the Kiyomura Corporation, owned by the Japanese businessman and restaurateur Kiyoshi "The Tuna King" Kimura. The company runs a popular chain of twenty-four-hour sushi restaurants, with branches near almost every major sightseeing location in Tokyo.

The Tsukiji Market in Tokyo, the largest wholesale fish and seafood market in the world. In 2018, the inner wholesale market was relocated to Toyosu Market, along Tokyo Bay, but the outer retail market and restaurants remain a major tourist attraction. (Image via Wikimedia Commons.)

Bluefin tuna is highly prized for its exceptional taste and texture. These fish typically sell for up to $40 per pound, but the price can rise to over $200 per pound depending on the time of year and availability. This particular bluefin was of such exceptional quality and texture that it commanded nearly $5,000 per pound.

Bluefin tuna's increasing popularity over the decades has resulted in rampant overfishing. The IUCN's Red List of Threatened Species lists Pacific bluefin tuna as "near threatened," with current stocks of the species estimated at near historical lows but not yet at immediate risk of extinction. The modern economics of global commercial fishing has compounded the problem. You might think that as demand for highly valued species increases, the cost to find, capture, and bring them to market would become so high and the market prices so exorbitant that bluefin tuna would be replaced on menus with more readily available species. This is not the case. One reason is that governments around the world spend billions of dollars every year to subsidize commercial fishing, in order to keep fish prices low and fishing communities viable. While well intentioned, these subsidies result in the depletion of fish stocks, lower yields, and depressed wages for fishers. A second reason is that increasing market prices for bluefin tuna and other overfished species do not prevent overfishing. Rather, they often drive a frenzy of overexploitation as fishers try to cash in while they can and maximize their returns. This is a classic tragedy of the commons scenario: people become unwilling or unable to cooperate to preserve the resource for the common good and for future fisheries. Scarcity and exclusivity also fuel consumer desire for luxury meals: eating bluefin tuna or other threatened species becomes a sign of wealth and social status. Overfishing has also accelerated as more people rely on fish as their primary source of protein, a trend linked to both population growth and the increasing global popularity of seafood.

Coupled with increasing demand for fish, some commercial fishing practices, like the use of bottom trawlers and supertrawlers, are particularly destructive to marine life. Bottom trawlers employ weighted nets dragged across the sea floor. Supertrawlers drag floating nets ranging in length from three hundred feet to seven miles. These techniques can destroy entire ocean habitats and capture a variety of unintended bycatch, such as sea turtles, sharks, albatross, manta rays, and dolphins, which are thrown back into the ocean, often dying or already dead.

Fishing is not inherently bad for the oceans. The problem comes when we harvest fish faster than the stocks can replenish themselves. We often don't recognize or respond to the gradual depletion of fish stocks until populations have been fished to critically low levels. All told, overfishing and habitat destruction have

depleted worldwide fish stocks by at least one-third. Where commercial fishing is particularly intensive, the situation is even worse.
In the Mediterranean Sea, for example, 90 percent of stocks are
overfished, and in European waters more generally, between 40 and
70 percent of fish stocks are now estimated to be at unsustainable
levels. Of all the threats facing our oceans today (among them climate change, pollution, ocean noise, plastics, and abandoned fishing gear), overfishing takes the greatest toll. We are systematically
harvesting ocean life to the brink of extinction. This threatens the
welfare of billions of people who rely on seafoods for sustenance and
the livelihoods of millions who make their living in some way connected to fishing.

In the late 1990s, an influential academic paper by the
University of British Columbia marine biologist Daniel Pauly and
colleagues attempted to explain the consequences of overfishing. The
authors described the effect as "fishing down marine food webs."
The authors argued that the mean trophic level of the global fishing
catch (its place in the food web) was declining. That is, humans have
tended to prefer large and long-lived species like tuna, billfish, cod,
and grouper, which mostly eat other fish. Having depleted stocks
of these large fish, we have systematically fished for lower-trophic-
level species such as herring, sardines, and anchovies. But today
even low-trophic-level shellfish like oysters and abalone command
exorbitant market prices because of their desirability and rarity
(caused by rampant overfishing). We eat and overexploit many different types of seafood, not just high-trophic-level species. While the
overall historical pattern may be one of fishing *down* the food web,
commercial fisheries have swelled to such a degree and demand for
seafood is so great that fishers often target multiple species—such
as tuna, cod, herring, and shrimp—based on the season, availability, market conditions, and other factors, so that we are now fishing

not only down but also *across* and *through* entire marine ecosystems. Understanding and addressing overfishing requires a holistic view of marine food webs.

One of the biggest challenges is that we lack information about the long-term health of oceans and fisheries. Systematic work by marine biologists to study, monitor, and collect data about the state of the oceans began only in the twentieth century, long after commercial fisheries had begun transforming stock sizes, marine food webs, and ecosystem structure. As early as the sixteenth century, European fishing fleets were intensively harvesting the Grand Banks of the western Atlantic for herring and cod. By the 1800s, an international trade in these fish—live or salted—was flourishing. Well before the Industrial Revolution, stock sizes were diminishing in areas near fishing communities. The effects of intensive fishing have been compounded by pollution, warming ocean temperatures, and other impacts. Modern fisheries data and historical data usually extend back no more than a century and have been compiled from marine systems already heavily altered by human activities. Fortunately, archaeology and deep historical data can provide information on the long-term health and structure of maritime fisheries and ecosystems. We can look to places where people fished and lived closely connected to the oceans for thousands of years.

INDIGENOUS FISHERIES ON CALIFORNIA'S NORTHERN CHANNEL ISLANDS

The Northern Channel Islands off the California coast, just north of Los Angeles, have been preserved and protected from much of the development that has transformed the mainland. Part of Channel Islands National Park, the four northern islands are, from east to west, Anacapa Island, Santa Cruz Island, Santa Rosa Island, and San Miguel Island. The islands have been separated from the mainland for well over one hundred thousand years and are home to animal and plant species found nowhere else on Earth. The Northern Channel Islands also contain thousands of archaeological sites that span at least thirteen thousand years of human history, back to the waning days of the last Ice Age—among the oldest and best-preserved archaeological records found anywhere in the world. The vast majority of these sites contain midden deposits—ancient refuse piles that contain discarded shells, animal bones, charcoal, artifacts, and other detritus left behind by the Chumash Indians and their ancestors. These middens are not worthless garbage but the stuff of archaeologists' dreams. The excavation and study of ancient refuse can tell us about the nature of local terrestrial and marine ecosystems; human hunting, fishing, and foraging; and the evolving nature of human-environmental interactions.

When Spanish explorers first sailed into the Santa Barbara Channel in 1542, they encountered large, sedentary populations of Chumash Indians living in thriving coastal towns and villages on the mainland and the islands. The first Spanish explorer, Juan Rodríguez Cabrillo, and others who followed described the intensive maritime fishing of the Chumash, whose fishing technologies included redwood plank canoes, woven fishnets, and shell fishhooks. Archaeologists once believed the Chumash maritime economy developed only in the past few thousand years, but we now know that the Channel Islands were first settled by maritime peoples at least thirteen millennia ago, employing coastal subsistence strategies that survived into the modern age. Along with a number of other Chumash groups that are not officially recognized by the US federal government, the largest Chumash tribe, the Santa Ynez Band of the Chumash Indians, continues its cultural, language,

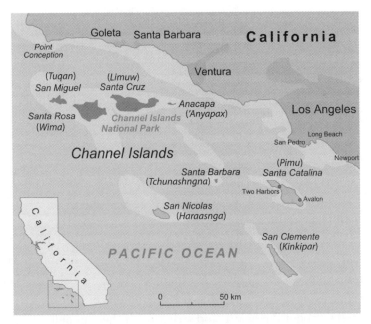

Map of coastal southern California and the Northern and Southern Channel Islands. Native American names for each of the Northern (Chumash Indians) and Southern (Gabrieleño-Tongva) islands are in parentheses. (Image via Wikimedia Commons.)

and ancestral traditions today. Hundreds of its members live on the Santa Ynez Reservation north of Santa Barbara.

Living in sedentary communities is unusual for hunter-gatherer groups. Most were nomadic or seminomadic, moving from one place to another to access seasonally available resources and avoid overharvesting in any given area. In most regions, the shift to permanently occupied villages did not happen until the advent of agriculture. The Chumash, along with a few other Indigenous groups in the Pacific Northwest, southern Florida, and elsewhere in the world, were able to harvest enough resources from their surroundings to support permanent settlements. Communities were led by hereditary chiefly leaders, who handled social relationships,

131

oversaw important ceremonies and rituals, and managed hunting, fishing, and gathering rights. The densest Chumash populations occupied settlements around large and productive estuaries along the mainland coast, where plant foods such as acorns were abundant, along with animals such as deer, fish, and shellfish. The Island Chumash had few terrestrial food options other than blue dick plants, which produce beautiful blue-violet flowers and edible underground corms. For sustenance, islanders relied on marine resources and trade with mainland peoples. They had to shift and adjust their subsistence strategies as their communities grew and required more food and resources.

The archaeological records of the Island Chumash offer a historical case study of the interactions between humans and marine systems. Since the Chumash ate many of the same fish and shellfish species we do today, the discarded remnants of Chumash meals can tell us about the health and structure of these populations, the sizes and densities of these species, and the relative stability of food webs in the past. We can compare these to modern fish stocks. Such information can give us a millennia-long perspective on the health of marine systems and the effects of commercial fishing on modern fish stocks and on human communities.

MODERN LESSONS FROM ANCIENT FISHING

When modeling ancient subsistence patterns, archaeologists often assume that hunter-gatherers targeted large species first and turned to smaller species only when the larger ones were unavailable. This assumption grows out of evolutionary theory and postulates that humans developed strategies to maximize caloric returns. By this reasoning, since large-bodied animals provide the most calories for the energy expended in hunting them, the best use of your time as a hunter-gatherer is to go after large animals first. In terrestrial systems, the largest animals tend to be lower-trophic-level herbivores, which also tend to be more abundant than higher-trophic-level carnivores. From a resource-conservation standpoint, this makes ecological sense and can be a fairly stable adaptation in terrestrial systems. In marine ecosystems, on the other hand, the largest animals tend to be the higher-trophic-level ones—such as sharks, whales, sea mammals, and tuna—which are often less abundant and more difficult to capture than smaller, lower-trophic-level animals. Many are found in the open sea rather than intertidal zones, and hunting them may require seaworthy boats and specialized hunting and fishing equipment (such as harpoons, nets, and fishhooks).

When European explorers first arrived in the Santa Barbara Channel, they found resource abundance rather than depletion, and they were astonished by the number of Chumash residents and the scale and bounty of the marine resources they harvested. One question that my colleagues and I have asked is how the Chumash and their ancestors avoided depleting marine systems and local resources as modern commercial fisheries have done today. Understanding how the Chumash accomplished this, despite an expanding population and natural climatic fluctuations in sea surface temperatures and marine productivity that affected fish and shellfish stocks, may offer insights into how we might sustainably harvest the oceans today.

Middens on the islands dating back eight thousand years or more are dominated by the discarded remains of shellfish. Dietary reconstructions suggest that shellfish provided about 90 percent of the animal protein in Chumash meals, with the remaining 10 percent coming from fish, birds, and sea mammals. Archaeologists believe that low-trophic-level resources were the most important sources of protein for early islanders.

Within this general pattern, however, there may be some significant variation. Many of these early sites contain finely made projectile points and other chipped-stone hunting technology that were likely used for spearfishing and hunting birds and small sea mammals. Some early sites have also produced remnants of fishing tackle constructed from bird bones. At one famous archaeological site on San Miguel Island named Daisy Cave, thousands of fish bones have been recovered that date back more than ten thousand years, suggesting that kelp-forest fish contributed the bulk of dietary protein. It is possible that at other early sites, islanders were doing more fishing and hunting than dietary reconstructions suggest. Animals may have been butchered near island shorelines, or from boats, and their bones discarded before the hunter-fishers returned to their campsites and villages, effectively rendering these resources invisible to archaeologists. Still, looking at the overall picture across the islands, shellfish probably contributed the bulk of dietary protein at most early sites, and the first islanders likely relied heavily on intertidal shellfish for their daily meals.

The period from about eight thousand to three thousand years ago, which archaeologists call the Middle Holocene, was one of transition and change for the Chumash. Populations grew tremendously, and the first permanently occupied villages sprang up on the islands. Wealth and status differences arose among individuals and families, evidenced by some Chumash individuals buried with ornate shell and bone pendants and jewelry and others buried with few or no grave goods. This shift was likely the precursor to the chiefly leadership structure witnessed by the Spanish. With these changes came the need to extract more resources from the marine environment. Dietary reconstructions from sites dated to this interval show considerable diversity: while shellfishing was still the main source of animal protein, sea mammals and fish contributed increasingly high proportions. Research suggests that at many Middle Holocene sites, kelp-forest fish contributed 20 percent or more of the protein, and sea mammals such as sea otters and some cetaceans (whales, dolphins, and porpoises) contributed more than 10 percent.

As populations grew, the Chumash spent more time harvesting difficult-to-access sources such as fish and sea mammals (and some birds). To accomplish this, they devised new fishing tackle: curved, single-piece fishhooks and more diverse hunting toolkits were invented about 2,500 years ago. People also invested more time and energy creating the tools they needed to harvest local marine environments.

The most dramatic changes to Chumash social, economic, and subsistence systems occurred over the past two to three thousand years. During this interval, called the Late Holocene, the dietary shift to kelp-forest fish intensified, with shellfish contributing around 20 percent, sea mammals about 35 percent, and fish about 45 percent of dietary protein. Building on trends that began during the Middle Holocene, rapidly growing populations, the establishment of more villages, and harvest pressure on intertidal shellfish communities forced the Chumash to spend more time and energy in nearshore and kelp-forest fishing and sea-mammal hunting. They focused on low- to mid-trophic-level species such as rockfish, perch, and wrasses. High-trophic-level species of fish and sea mammals became more common prey only during the past 1,500 years. Fishing for species like tuna and swordfish, which was more dangerous and difficult, was likely tied to ceremonial feasts or ritual hunts; overall, these species contributed little protein to Chumash diets.

To facilitate fishing and sea-mammal hunting, the Chumash devised new technologies. One of the most innovative and important was large and stable redwood plank canoes, developed about

1,500 years ago. Not only did these boats offer greater access to deepwater fish and seal and sea lion communities on offshore rocks and islets, but they also facilitated trade and communication between the islands and the mainland. Bow-and-arrow technology also first appeared around this time and improved the efficiency of sea-mammal hunting. The Island Chumash also accelerated the production of shell money beads, which were used to pay for goods and services throughout Chumash territory (and beyond). These changes resulted in rapid shifts in Chumash society, such as greater cultural complexity, craft specialization, and elite control, and they gave rise to the Chumash lifeways first documented in the sixteenth century by early Spanish explorers.

Throughout their history, Island Chumash subsistence strategies, for the most part, ran counter to global historical patterns. Rather than fishing down marine food webs, the Chumash fished *up* food webs, focusing first on low-trophic-level shellfish, then turning increasingly to kelp-forest fish and sea mammals as population densities increased. Although some of the earliest sites on the Northern Channel Islands show that the Chumash were fishing for local kelp-forest and nearshore species, capturing birds, and hunting seals, these were probably not the basis of their diet. Rather, Channel Islanders focused on readily available intertidal shellfish, which were easy to capture and prepare.

Both the islands and the mainland were dotted with Chumash villages, all of which relied heavily on maritime fishing, hunting, and foraging year-round. For thousands of years, Chumash hunters and fishers relied on the same watersheds. We would expect them to have had significant impacts on local ecosystems, especially over the past three thousand years, when population densities and maritime exploitation were highest. So how did the Island Chumash maintain relatively sustainable fisheries over the long term? New interdisciplinary methods in archaeology and ecology and the chemical sciences are helping us answer this question.

FISHING ACROSS MARINE FOOD WEBS

At the far eastern edge of Santa Rosa Island lie the remains of one of the largest and most important Chumash villages on the Northern Channel Islands. To the untrained eye, not much is left to see. A decommissioned ranching and National Park Service road runs just to the west. If you head toward the coast right after crossing Old Ranch Canyon, the largest drainage on the island, you've arrived. The village of Qshiwqshiw, which means "bird droppings" in the Chumash language, consists of two archaeological sites positioned right along the coast, one to the north of Old Ranch Canyon and a larger, more impressive one to the south. Much of the southern site is buried below a sea of introduced grasses on a low-lying terrace flanked by beautiful sandy beaches to the east and a freshwater marsh to the north. Heading out onto the coastal terrace, you need to watch where you step, as at least nine circular depressions are concealed by the knee-high grass. These shallow, pit-like features, several meters in diameter, were once the semisubterranean dwellings of Qshiwqshiw residents. The houses (or 'ap) would have had roofs made of sea-mammal skins or cattails and other vegetation, secured with bent poles and branches.

Archaeologists have long known of Qshiwqshiw. The Chumash elder Juan Esteban Pico first described the site to anthropologists in the 1880s. Periodic archaeological investigations in the 1960s, '80s, and '90s suggested that the northern site was the epicenter of village life after Spanish contact. Directly behind the beach in this area is a large scatter of discarded shell, fish-bone, and stone-tool debris, along with eight house depressions. From these deposits

137

archaeologists also identified glass trade beads and shell beads drilled with metal needles (rather than stone drills), both hallmarks of the period following European contact. Despite more extensive archaeological excavations, sampling, and radiocarbon dating at the southern locus than the northern, no clearly postcontact artifacts have been identified in this part of the village. The location was likely abandoned at or shortly after European arrival, perhaps because of a shrinking population ravaged by the introduction of deadly diseases.

In the summer of 2012, I directed a team of archaeologists and Chumash consultants in small-scale excavations of the southern locus of Qshiwqshiw. We were interested in better understanding the ways in which the Chumash harvested resources and whether the harvesting was done sustainably. Qshiwqshiw was one of the largest villages on the Northern Channel Islands, with four Chumash chiefs and at least 119 residents, based on baptismal records. These records, compiled by the Spanish missions from the mid-eighteenth to the mid-nineteenth century, offer a reliable minimum estimate of population size for much of the Spanish territory in early California. (Such data sources do not account for the fact that not all Indigenous peoples were baptized; nor do they reflect the population losses to disease and disruption to traditional lifeways wrought by Spanish colonialism.) This was a time when at least three thousand people lived on the Northern Channel Islands. Fishing was restricted to watersheds directly adjacent to the villages so as to not encroach on the foraging territories of adjacent villages. Especially with a village the size of Qshiwqshiw, this limitation would have put great stress on local resources. We were interested in whether human-induced stresses could be detected from the village refuse deposits.

As expected, fish was the main source of dietary protein at Qshiwqshiw. Like residents of other large villages across the islands,

Qshiwqshiw residents favored kelp-forest fish, which required considerable investments of fishing technology (such as tackle, boats, and lines) and time. Intertidal shellfish species such as California mussels, marine snails, and abalone were easy to gather and process, but these resources were inadequate to meet the villagers' needs. Measurements of mussel and abalone shells from Qshiwqshiw and other large Chumash villages demonstrate that intensive human harvest of these resources over many hundreds of years resulted in a reduction in their overall sizes, a trend that would have propelled fishers to focus more on fish. The most common fish captured at Qshiwqshiw were surfperches such as pile perch, followed by rockfish, skates and rays, and then houndsharks. Based on the ecological roles of these species today, it seems that the people of Qshiwqshiw heavily fished kelp-forest fauna, particularly species from middle trophic levels, but also consumed species from other habitats and from both higher and lower trophic levels.

Along with identifying the food remains at Qshiwqshiw, we used chemical analyses of carbon, nitrogen, and amino acids in the fish bones to reconstruct local food webs and gauge the health of the marine ecosystem. We found that prior to Spanish arrival, pelagic (offshore) and nearshore environments were energetically coupled, meaning that they supported one another to create stable and healthy marine ecosystems. Just as a diversified stock portfolio protects the owner from the failure of a single company or type of asset, plant and animal communities that are supported by multiple sources of energy from multiple habitats can overcome short-term impacts or disturbances caused by natural climatic change or intensive human harvest. Our data suggest that the marine systems of the Northern Channel Islands before Spanish contact were well-balanced and resilient, with nearshore, kelp-forest, and offshore environments all supporting one another.

This balance was a critical factor in sustaining intensive Chumash fisheries. Even as Chumash fishers ramped up their harvest of local resources, their impacts on food webs were likely balanced by nutrients and energy flow from adjacent systems. Although Chumash harvest strategies focused on mid-trophic-level kelp-forest fish, Chumash fishers also captured prey from diverse environments and trophic levels. Rather than fishing up or down marine food webs, they fished across a range of marine food webs. This practice stands in stark contrast to postindustrial fishing, which tends to focus on harvesting a single species or single habitat to unsustainable levels. Once one species or habitat is overexploited, commercial fisheries simply target new ones. Such a strategy severs the energy links between environments and ecosystems that enable these systems to withstand stresses.

REGULATING CATCHES

Considering the entire marine system from a long-term, holistic perspective is an important part of marine conservation. Archaeological investigations of the history of the Chumash and their ancestors on the Northern Channel Islands going back more than thirteen thousand years, along with detailed analyses of faunal remains from Chumash villages such as Qshiwqshiw, provide insights about how ancient marine systems evolved into the ones we know today. An ecosystem approach to environmental ethics and conservation recognizes that individual organisms and species are parts of a whole, integrated system that cannot be effectively managed without acknowledging its interconnectedness. Yet management decisions that affect the daily lives of people and animals are often made and enforced at the species level. Fishers today are required to follow rules about when and where they can fish, the technology or tools they are allowed to use, and what quantity and types of fish they can keep. The goal is to maintain fish populations that remain stable and healthy despite human harvest and environmental fluctuations. Mechanisms such as fishing licenses, limited fishing seasons, and size and catch limits are some of the most common and effective ways that fishery biologists build and maintain healthy, sustainable fish stocks.

Like other fisheries management policies, however, modern fishing laws and regulations are typically based on modern or shallow historical records, rarely extending back more than one hundred years or before the start of extensive commercial fishing. This limited perspective can lead to regulations that manage toward dysfunction, such as size and catch limits based on fish populations already heavily affected by commercial exploitation. It is impossible to know what the average fish size should be or what a healthy fish population should look like if commercial harvesting has already taken a heavy toll.

Recent research suggests that establishing fishing baselines built from deep historical data and implementing management practices that preserve larger (older) fish can be effective in combating the shifting baselines syndrome (see chapter 1). Large fish, which tend to produce more offspring, are critical for maintaining healthy population sizes. If the baseline for a healthy fish population is set using data from an already heavily fished and overexploited population, size limits may be set far too small, at levels that slowly erode the population as the younger fish are unable to reproduce at a rate that outpaces harvest.

Setting sustainable size limits, however, requires knowing what constitutes a large fish. If heavy fishing pressure has targeted large fish for decades to centuries, the largest fish found today may be much smaller (and younger) than they were in the deep past. Fish bones from archaeological sites offer an answer. For over a decade now, colleagues and I have used fish-bone remains from Chumash archaeological sites to estimate the sizes of prey fish over thousands of years in southern California. We begin by measuring modern fish from southern California and fish bones from museum collections. For each species of interest, we measure individual bones—mostly from the head—from dozens of fish of known lengths. From these standardized measurements, we develop regression formulas that can accurately determine the total length of the fish from measurements of specific bones. When we have identified archaeological fish bones by species, we measure them and plug the measurements into our regression formula to gauge the size of the fish.

Rockfishes and California sheephead are species of particular interest. Rockfishes are a diverse group of finfish comprising over one hundred species. They come in a dizzying array of colors and sizes and inhabit nearly every coastal habitat along the northeast Pacific. Some species can live more than two hundred years. Coastal Native American communities such as the Chumash have harvested rockfishes for thousands of years, and, under the more palatable moniker of Pacific red snapper, they remain popular on restaurant menus and dinner plates. The commercial harvesting of rockfish began along the central California coast in the mid-nineteenth century, expanded in the 1960s, and grew into one of the most profitable fisheries in North America. In the past several decades, however, overfishing, warming ocean temperatures, and habitat deterioration have resulted in alarming stock declines of nearly every rockfish species, especially in southern California. Fisheries managers have enacted widespread closures of the commercial, sport, and recreational fishery in an attempt to curb the decline in rockfish biomass.

The California sheephead (*Semicossyphus pulcher*) is a distinctive fish found along the eastern Pacific coast from Monterey Bay, California, to Cabo San Lucas, Mexico. California sheephead inhabit

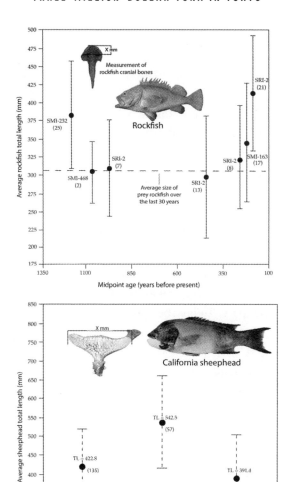

Modern fish sizes compared with those captured in the ancient past on the California Northern Channel Islands. Top: Rockfish today are on average 50 millimeters (2 inches) or 14 percent smaller than those captured over the past 1,400 years by Chumash fishers. Bottom: California sheephead today are 70 millimeters (2.8 inches) or 15 percent smaller than sheephead captured over the past 3,000 years. (All images via Wikimedia Commons or by author.)

nearshore rocky reefs and kelp forests and can live for twenty to thirty years. With powerful jaws, sharp teeth, and a rigid throat plate for grinding shells, they prey on sea urchins, crabs, lobsters, and other shellfish. They are hermaphrodites: all sheephead are born female and capable of changing into males when environmental conditions, social cues, or limited availability of males trigger them to do so. These fish are critical to the health of California kelp-forest ecosystems, as their predation of sea urchins limits the deforestation of kelp.

Like rockfish, California sheephead have been fished by coastal Indigenous communities for millennia. Heavy commercial exploitation in southern California began in the late 1980s, triggered by the demand for live seafood both domestic and abroad, especially in Asia. By the 1990s, the commercial sheephead fishery had quadrupled, and the recreational and sport fishing for sheephead spiked in popularity. Today the sport fishery exceeds the commercial one, and southern Californian populations are considered vulnerable by the IUCN.

One of the challenges of managing hermaphrodite fish species is that fisheries target the largest individuals, which tend to be males. In response to the lower number of males in the population, increasingly smaller and younger females convert to males. This process results in a decline in the number of eggs produced by females, leading to overall population reductions.

Using measurements of rockfish and sheephead fish bones recovered from Northern Channel Island archaeological sites, we compared sizes of fish captured in the past with those captured today. We found stark differences. Modern rockfish are nearly 50 millimeters (2 inches) smaller than the average rockfish captured by the Chumash—a 14 percent reduction. These findings suggest that even the intensive fisheries of the Island Chumash captured rockfish that were, on average, larger and older than those today, even though the Chumash primarily fished nearshore populations that were likely younger and smaller than the deepwater rockfish targeted by the modern commercial and sport fishery.

A very similar pattern was identified with California sheephead. Modern sheephead were nearly 70 millimeters (2.8 inches) smaller than sheephead found at Northern Channel Island archaeological sites over the last three thousand years—a 15 percent reduction in the average size. Again, particularly given the limitations of Chumash fishing technologies, this pattern is alarming. Rockfish and sheephead populations in southern California in the deep past commonly included considerably older, larger individuals than we find today.

Although the average size of prey fish was significantly larger in the past, midden deposits often contain the remains of fish of a great range of sizes from the same species. Intentional or not, capturing fish of different sizes may have helped keep fish stocks healthier. The survival of the largest fish results in the production of more and larger eggs, which leads to more resilient fish stocks. These findings suggest that it may be time to rethink minimum size limits for many commercially and recreationally important fish species. A strategy that resembles Chumash fishing practices—one that is based on hook-and-line fishing, targets a broad array of fish sizes, allows for refugia populations (such as the deep waters that Chumash fishers could not access), and maintains a balanced population structure of young and old or male and female fish—may be more sustainable.

SAVING ENDANGERED SEA LIFE AND
REBUILDING LOST FISHERIES

Ocean resource managers and marine biologists must often balance conflicting goals. Our oceans are reservoirs of biodiversity, indicators of climate change and planetary heath, and a resource that sustains human and nonhuman communities. These interests can sometimes lead to disputes over which marine species to protect and whether their protection threatens the viability of maritime economies and fisheries.

This debate over marine conservation is skewed by historical circumstances. First, because scientific datasets on ocean health date back no more than a century, long after commercial fisheries had begun taking a heavy toll on fish stocks, we need alternative data sources to understand the "original" state of the oceans. Second, the hyperproductivity of many early commercial fisheries has set up unrealistic expectations for modern enterprises. A return to similar levels of exploitation is neither sustainable nor healthy. The ultimate goal for marine conservation and fisheries must be to restore resilient marine habitats and systems and develop sustainable fishing practices that operate as an integrated part of these systems. Once again, perspectives from archaeology, deep history, and Indigenous fisheries can guide this process.

One excellent example comes from the eight-thousand-year history of California red abalone fishing on the Northern Channel Islands. Red abalone is a large, slow-growing marine mollusk that thrives in cold waters and is found in rocky, kelp-forested areas in the low intertidal to subtidal zones from Oregon to Baja California. The Chumash harvest was intensive and continuous, but there is little evidence of overfishing of this abalone during ancient times. Commercial harvesting began with Chinese immigrant fishers about 170 years ago. In the 1960s, the Pacific red abalone was the focus of a multimillion-dollar industry. By the 1970s and 1980s, serial overfishing, increased competition with recovering sea otter populations, and withering syndrome (a bacterial disease) took a dramatic toll on abalone populations up and down the North American Pacific

coast. In 1997 a moratorium was placed on all California red abalone fishing south of the San Francisco Bay, leaving open only a highly
regulated sport fishery in northern California.

Red abalone was the last of California's abalone species to be
closed to commercial and sport exploitation, as moratoriums had
been placed on the fishing of black, green, pink, and white abalone
varieties by the California Fish and Game Commission in the early
1990s. Despite careful management, restoration efforts, and nearly
two decades of fishery closures, there have been few or no signs of
improvement for most of these species. The commercial harvest of
abalone has evaporated in California, and the outlook for its return
is abysmal.

Red abalone may be the one bright spot for the recovery, restoration, and management of the California abalone fishery. They
have expanded their numbers and range across much of southern
California, especially along the Northern Channel Islands, likely
spurred by the strong upwelling and cold-water influx that made
these waters rich in nutrients for marine life and the focus of
commercial and recreational harvests in the twentieth century. An
amendment of the California Abalone Recovery and Management
Plan allows for experimental harvests before stocks are fully recovered, and commercial abalone divers have argued for over a decade
that the California Department of Fish and Game should open a
test fishery along San Miguel Island and allow them to return to
their lost livelihoods.

Debates over this proposal are especially contentious. Commercial fishers see an opportunity to revive an economically important and iconic fishery, while many marine biologists and resource
managers are concerned that an experimental fishery might negate
the hard-fought progress toward full red abalone recovery. At the
same time, marine mammal conservation groups are campaigning

for the reintroduction of sea otters in the Northern Channel Islands and elsewhere along the southern California coast. These charismatic marine mammals were nearly driven to extinction by fur hunters in the mid-1700s but have recently expanded their range and numbers throughout Californian waters. The trouble is that sea otters are voracious predators of red abalone and other shellfish, and commercial fishermen are concerned that their expansion will harm the recovery of red abalone and the viability of the contemporary commercial sea-urchin fishery, especially along the Northern Channel Islands. This morass of competing interests and competing priorities involves not only a struggle between commercial fishers and conservationists but also disputes over which marine organisms should take priority in conservation.

One way to find a solution is to try to understand how the Island Chumash managed a thriving fishery for abalone, sea urchins, and other shellfish for millennia in an area where sea otter communities were present. In recent years, colleagues and I have synthesized all of the archaeological evidence for red abalone fishing on the Northern Channel Islands. Integrating paleoecological, historical, and modern catch data, we found patterns indicating that the availability of red abalone was discontinuous across the Santa Barbara Channel. Archeological, historical, and ecological data suggest that beginning about eight thousand years ago, Chumash hunters reduced sea otter populations in local watersheds, either by hunting or by scaring otters from local watersheds. During this interval, red abalone grew so large that they could not fit in the cracks and crevices where they typically hid from sea otter predators. Marine biologists assure us that red abalone would never reach these sizes in a kelp-forest ecosystem with abundant sea otter predators, a pattern confirmed by the commercial fishery data postdating the local extirpation of sea otters from southern

California. This resulted in exceptionally productive red abalone fisheries (as evidenced by archaeological shell middens packed full of large red abalone shells). There is no evidence, however, that sea otters were eradicated from the Santa Barbara Channel before the arrival of Euro-American commercial fur hunters. The reduction of sea-otter predation on abalone, mussels, sea urchins, and other shellfish bolstered Chumash harvests despite intensive human predation pressure. Marine food webs were restructured, with humans replacing sea otters as top shellfish predators. This was particularly true around San Miguel Island, where red abalone were fished for ten thousand years, regardless of fluctuations in local water temperatures. However, they were abundant on the other islands only during optimal climatic conditions, when sea surface temperatures were colder than normal.

This deep history teaches us two important lessons about red abalone abundance and fisheries in southern California. First, the fact that San Miguel Island waters sustained intensive red abalone fisheries for thousands of years, regardless of changing climatic conditions, suggests that these waters are a critical habitat for red abalone larval production and recruitment. Commercial and sport fishermen, resource managers, and marine biologists should expect red abalone communities to recover round San Miguel Island first. Their recovery along island shorelines to the east would offer a measure of their population rebound and ability to sustain a test fishery. If red abalone recover in these secondary locations, Channel Island red abalone communities should be able to sustain some reasonable level of human harvest pressure. Second, history tells us that a red abalone fishery and sea otter communities can coexist in southern California. Red abalone populations must be fully recovered before a sustainable human harvest is viable, and sea otter populations would need to be carefully reintroduced and their populations controlled. Ultimately, the recovery of red abalone would re-create an ecological state that persisted for millennia on the Northern Channel Islands. If we are willing to listen, the past can help us repair the ecological damage done by both the overfishing of abalone and the overhunting of sea mammals in California.

FUTURE OCEANS

The future of ocean ecosystems hangs in the balance. Business as usual is likely to result in the continued reduction of biodiversity, the collapse of economically critical fisheries, and the degradation of marine habitats. The impact of human activity on the vast, diverse, and complex systems of the oceans is often hard to see or understand. About 50 percent of the Earth's surface consists of ocean waters outside the boundaries of any national territory. These areas are therefore subject to very few regulations. No single nation or government is looking out for their well-being; protection efforts are often undertaken by nongovernmental organizations. In addition, many of us do not have a direct connection with the ocean or marine life. Growing up in the American heartland, I had the opportunity to see, enjoy, and appreciate an ocean only during spring breaks in Florida. It was not until I graduated from college and started my first "real" job in Portland, Oregon, that I lived within driving distance of an ocean and started to feel more connected to the mysterious, briny seas.

Yet the oceans are critical for all of humanity, even those who will never see them. Ocean currents regulate weather and temperature around the globe. Healthy oceans reduce human climate-change impacts by absorbing over 90 percent of the heat and 30 percent of the carbon dioxide produced annually by human actions. Tiny ocean-dwelling phytoplanktons are responsible for over 50 percent of Earth's oxygen production. The world's fisheries act as a cornerstone of human subsistence. Fish provide 17 percent of all the animal protein consumed globally, and over 25 percent in developing countries. Fish are the single most traded food commodity on Earth, and fisheries account for some two hundred million jobs. With an estimated 90 percent of wild fish stocks at or below sustainable capacity, the time to preserve them is now.

It can't be business as usual. The ways in which we have studied, managed, and tried to protect the oceans up until now have failed. We need bold new thinking to avoid making the same mistakes. We must look to the past to establish management baselines and provide more informed measures of ocean health. Rather than focus on sustaining fisheries or maintaining current yields, we must concentrate on rebuilding fisheries and marine habitats. We know that local, regional, and global fisheries and marine habitats have been severely depleted and degraded. Hyperabundant historical and modern fish and shellfish catches are anomalous: they are predicated on ecosystem dysfunction, rampant overharvesting, and unrelenting profiteering.

There are many steps we can take to help the oceans. Along with reducing our carbon footprint and energy consumption, we can make sustainable seafood choices when grocery shopping and dining out. The Seafood Watch website and phone app maintained by the Monterey Bay Aquarium help North American consumers make informed decisions about what types of seafood to buy (seefoodwatch.org). When we make these choices consistently, we spark change throughout the supply chain, pushing companies to improve their fishing and aquaculture practices. We can also support the many institutions and organizations working to protect the ocean, financially or by volunteering. Examples include Ocean Conservation (oceanconservancy.org), Oceana (oceana.org), and

Surfrider Foundation (surfrider.org). Colleges and universities, even far inland, often have professors, institutes, and students working on issues of ocean conservation. Public aquariums like the Monterey Bay Aquarium, the New England Aquarium, the Audubon Aquarium of the Americas, the Shedd Aquarium, the California Academy of Sciences, and the New York Aquarium (among others) are leaders in marine conservation and research. They are great places to learn about our oceans, share concerns, and act to support them. And, of course, we need government action. We can research the ocean policies of public officials before we vote, and contact local officials to let them know we support marine conservation projects. Before politicians can support the issues that matter to us; they must know our concerns. Be heard.

●

PLANET EARTH IN THE AGE OF HUMANS

D espite the very good advice to live in the moment, we must all look toward the past. As a professional archaeologist, I have lived much of my adult life embedded in the past, discovering and decoding stories from the past and looking for answers to questions about human trajectories. But history and the past are important for us all. History, in all its forms, is essentially the accumulated wisdom of all people through the ages. It is the vast body of facts, experiences, skills, and information we call knowledge, passed from one generation to the next. All of us rely on historical knowledge every day. To try to make the best choices, we consult our lived experiences, lessons learned from our elders, and the stories we have heard. History broadens our outlook on the world and helps us confront new situations and challenges. History helps us make informed decisions and recognize the influence of biases, emotions, and memories. For all these reasons and more, history matters.

Nevertheless, we have a tendency to disregard history when contemplating environmental, ecological, and evolutionary change. Ecologists, biologists, chemists, climatologists, and other scientists tend to be less engaged with the influence of humans and less interested in historical trajectories than in material evidence and modern experiments and field investigations. This does not mean, however, that these scientists dismiss history out of hand or discount the importance of long-term human-environmental interactions. Integrating historical data into modern management plans can be challenging. These data are often very different from those collected by scientists who study modern ecosystems, climates, and biota. In most cases, archaeologists and other historical scientists cannot offer data that can be quickly and easily added to an ecological or climatological model. Rather, the study of history offers narratives and perspectives that can help us chart a better path forward.

Still, history has much to offer if we are willing to listen and think creatively about its application. It tells us about the potential power of reforesting large tracts of land across the globe and suggests how better land-management strategies might reduce greenhouse gases and fight global warming. It offers clues about how threatened plants and animals once thrived in local environments and how their biogeographies may have changed as a result of human hunting, anthropogenic climate change, and other modern impacts. For many species that have suffered sudden and dramatic decline, we have little recorded information about their natural abundances and distributions. Archaeology and other deep historical information become critical sources of guidance. Histories of Indigenous communities can help us reconstruct the environments of the ancient past and tell us how these groups developed resilient ecological strategies that sustained them over millennia. These narratives and the lessons from history are just as powerful as the charts, projections, and simulations often produced by modern ecological and climate scientists—or more so.

Even though the Anthropocene as a scientific and historical concept is still being debated, I find it hard to believe that any reasonable, open-minded, and observant person could look at the world around them and not agree that we are reshaping the planet in our own image. Mining, dredging, fracking; constructing roads, houses, skyscrapers, and cities; plowing fields, clearing forests, constructing pipelines, and rerouting waterways; pollution, landfills, commercial agriculture and fishing, and many other human activities are reshaping the character of land and sea. We are modifying the chemistry of our oceans and atmosphere so dramatically that we have to delve deep into Earth's geological past to find evidence of similar changes. Humans now emit somewhere on the order of 34 billion tons of atmospheric carbon dioxide every year. We would need to travel back in time 2.5 billion years to find a similar disruption to the natural cycling of carbon on our planet. Any way you slice it, humans are changing the planet.

We have been talking for decades about humanity's impact on the Earth and the potentially dire consequences of our collective actions. Rachel Carson's *Silent Spring*, published in 1962, is an enduring classic, often cited as the one of the best science books ever written. Carson was one of the first writers to raise public awareness about the health and viability of our world's flora, fauna, and ecosystems, and to expose the links between pollution, pesticides, human impacts on nature, and public health. It's been over fifty years since the first Earth Day, on April 22, 1970, which marks the effective launch of the modern environmental moment. Since that time, mountains of academic articles, popular and esoteric books, magazine and newspaper articles, poetry, films, television shows, art, speeches, musical concerts, plays, and virtually every other form of human communication have told the story of what we are doing to the planet, most of them pleading for us to find a better way forward before it is too late.

The concept of the Anthropocene was first proposed in 2001 to draw attention to the accelerating environmental crisis created by humans. Perhaps for the first time, the vast majority of people—academics, students, environmentalists, concerned citizens, young, old, poor, rich—now believe that we live on an imperiled Earth and that humans are to blame. We are no longer focused on what we *may* be doing to the planet and what the consequences of these actions *might* be. The majority of our global community recognizes that we have fundamentally altered our planet and need to address the problem, now. The consequences of our decision-making and actions have arrived. The question is what, if anything, we can do.

The notion that we are living in a world utterly transformed by human action has resonated with broad scientific, scholarly, and public communities. The Anthropocene idea has crossed academic disciplinary boundaries, appearing not only in geological and ecological scholarship but also in papers by philosophers, artists, anthropologists, and chemists. It has facilitated communication across disciplines and created new and interesting conversations about the state of the world. While its definition and connotations may vary among disciplines, the Anthropocene has created meaningful and productive areas of overlap for many researchers who might otherwise struggle to find common ground.

The concept of the Anthropocene has even entered public discourse. Prominent and widely read magazines such as the *Atlantic*, *Smithsonian* magazine, *Newsweek*, and *Discover* have run stories about the Anthropocene—what it is, what it means, and why people should care. One of my favorite examples was published in a 2011 issue of the *Economist*. The cover image showed an Erector set–like Earth blowing several gaskets and a headline declaring, "Welcome to the Anthropocene." The idea offers scientists, conservationists, resource managers, and many others working on issues of sustainability and environmental management a way to connect with the public and communicate more effectively about some of the most important issues facing the global community.

It is the Anthropocene concept, the public interest it has generated, and over ten years of writing and thinking about these ideas that have moved me to write this book and given me a platform for demonstrating why archaeology and what I do matter for the modern world. Lots of people from all walks of life and professions are intrigued by archaeologists, what we do, what we dig up, and how we learn about the past. We have a captive audience of people who want to hear what we have to say. (On the other hand, some folks are simply bemused when they meet a professional archaeologist and quip, "Someone pays you for that?" My response is usually "Yep, but not much.") It's fun to learn about the history and archaeology of ancient Egypt, Mesopotamia, Mesoamerica, and other places. The problem is that many people view archaeology as entertainment rather than an intellectual pursuit with practical applications for their daily lives. That is where the Anthropocene comes in.

The Anthropocene has created a global conversation that shows how human actions have created the modern environmental crisis. Environmental calamity is here, we created it, and we must do something about it. And if we are to find solutions, the past must be our guide. In this book I have tried to show that history offers answers to critical questions about how we brought about the Anthropocene, what the world looked like before significant human impacts, and how we might address the problems we have created and bring about a more sustainable future. History is of vital, practical environmental importance in an Anthropocene future. How can we know what effective greenhouse-gas targets look like without ice cores that tell us about the Earth's atmosphere in the past? How will we know where forests once existed and what plant and animal species thrived in them without the archaeological analysis of botanical remains and animal bones, without pollen cores, without historical texts? How can we save endangered plants and animals without understanding their evolutionary and biological histories, the conditions under which they once thrived, and the long-term causes of their population declines? How are we going to save the oceans without knowing what they looked like before commercial fishing, pollution, and ecosystem dysfunction, without understanding where and how many fish and sea mammals thrived in the past, and without identifying fish sizes in deep history? Archaeology and history are more than interesting pastimes or fodder for Hollywood scripts: they have practical importance for the future of the Earth.

Still, this idea is not universally shared. Attitudes need to change. My colleagues and I need to do a better job of educating the public and the highest levels of government about the importance of history in the Anthropocene.

Most of the environmental statutes put in place in the United States to protect human health and the environment were passed in the late 1960s through the early 1980s. Given current politics in the United States, it may seem ironic that it was a Republican, President Richard M. Nixon, who signed the National Environmental Policy Act and created the Environmental Protection Agency in 1970. The legislation ordered US government scientists and resource managers to include the "best scientific data available" when constructing environmental restoration and management plans. This requirement can be found in legislation such as the Marine Mammal Protection Act of 1972, the Endangered Species Act of 1973, and the Magnuson-Stevens Fishery Conservation and Management Act of 1976. Agencies have spent considerable time and effort defining what this somewhat vague mandate means and, in some cases, developing rigorous frameworks for determining what types of data their decisions should take into account. Typically these discussions center on the types of ecological and biological datasets that should be used. Rarely do data and perspectives from other disciplines even come under consideration.

Unsurprisingly, then, archaeology and deep history are not included. The lingering perception seems to be that the study of history is practically important only for avoiding the repetition of past mistakes. Often ignored are the ways that I believe history matters most. Archaeology and the historical sciences offer tools to analyze, contextualize, and explain the problems of the present and identify solutions, strategies, and patterns that might otherwise be invisible.

Jeremy Sabloff, an archaeologist, a member of the National Academy of Sciences, and a former professor at the University of Pennsylvania, advocates for what he calls "action archaeology." Sabloff is one of the most accomplished and best-known archaeologists in the world. For decades he has conducted fieldwork and research on ancient Mayan civilizations. He has also been one of the most vocal proponents of the relevance of archaeology to the modern world and the need for archaeologists to communicate its importance to the public. He argues that archaeological research must work with and serve living communities in diverse ways, from assisting city planners in designing sustainable cities to aiding Indigenous communities in land-claim cases. He also argues that politicians and policymakers should consult archaeologists and anthropologists, as they commonly consult economists, epidemiologists, and political scientists, in making important practical decisions.

Although the practical value of archaeological data may seem obvious, the use of deep historical data is sometimes contested and met with hostility even by academic scientists. Some scientists believe that its application to ecological restoration is impractical, driven by the fantastical goal of creating "living museums." These critics tend to dismiss historical perspective as useless "nostalgic recompositions of the past." Such lack of insight into the lessons offered by deep history makes the job of educating scientific and public communities difficult, but also critically important.

We all need to understand how and why the past matters for our future world. Even some of the most dramatic news stories about environmental change and the challenges we read about in local newspapers or see on network news shows—anthropogenic climate change, the fire at Notre-Dame, the global extinction of iconic plants and animals, and the crisis of our oceans and overfishing—have deep historical roots. Understanding their causes can help us build sustainable solutions.

Addressing the environmental challenges of the present will require conversations that include ecologists, biologists, historians, archaeologists, social scientists, and other specialists. These conversations must also include the voices of community stakeholders and Indigenous groups. In the past the environmental movement has often neglected or actively dismissed the knowledge and experiences of these diverse groups. Just as history is important for addressing our environmental challenges, the perspectives and knowledge offered by marginalized groups can provide valuable insights and key components of successful environmental management plans.

We face an uncertain future. While we are certain that we are living through an environmental crisis of global proportions and that this crisis has been created by humans, it is unclear what the outcomes will be. We lack clarity on the future direction of climate policies, greenhouse-gas emissions, complex interconnected climatic and socioeconomic feedback loops, rates of plant and animal extinctions, or environmental tipping points. We cannot allow this uncertainty to paralyze us. In science, there is always ambiguity, and the more complex the problem, the more uncertainty we must tolerate. Even with the best scientific techniques and information, the future of our planet remains uncertain and difficult to predict. We must embrace this uncertainty, realize that it is a constant and consistent part of our lives, and take bold action. By looking to the past, we can decode how the world once looked and how the blueprints and perspectives of history can help us move toward a more sustainable and resilient future. In the potential chaos and anxiety born of uncertainty, history can calm our emotions and help us build a better future for an imperiled Earth.

○

ACKNOWLEDGMENTS

This book has been a passion project six years in the making. Translating archaeological and deep historical research for a public audience has been an exceptionally challenging and rewarding undertaking, and I have many people to thank for their support, advice, collaboration, and friendship along the way. I have had the great fortune of working with an interdisciplinary group of brilliant scholars and have learned a tremendous amount from each of them. In particular, Ginta Ale'ke, Scott Anderson, David Ball, Linda Bentz, Seth Bruck, Mike Buckley, Breana Campbell, Cynthia Catton, Brendan Culleton, Loren Davis, Paul Dayton, Matt Edwards, Jon Erlandson, Elyssa Figari, Scott Fitzpatrick, Stephanie Gallanosa, Kristina Gill, Mike Glassow, Ken Gobalet, Lain Graham, Amy Gusick, Hannah Haas, Scott Hamilton, Marco Hatch, Courtney Hofman, Brian Holguin, Kristin Hoppa, Ann Huston, Dustin Kennedy, Lawrence Kiko, Laura Kirn, Shannon Klotsko, Cassie Krum, Thomas Leppard, Kent Lightfoot, Seth Mallios, Jillian Maloney, Joseph McCain, Matthew McCarthy, Iain McKechnie, Juliette Meling, Madonna Moss, Dan Muhs, Vijayanka Nair, Seth Newsome, Alex Nyers, Jenna Peterson, Anjali Phukan,

ACKNOWLEDGMENTS

Leslie Reeder-Myers, Laura Rogers-Bennett, Emma Elliott Smith, Kevin Smith, Paul Szpak, René Vellanoweth, Natasha Vokhshoori, Thomas Wake, Phillip Walker, Stephen Whitaker, and Lauren Willis helped shape my thinking and challenged me to be a better scientist, archaeologist, and educator. I owe special thanks to Torben Rick, who read multiple versions of *Understanding Imperiled Earth* and was always willing to give advice and help along the way. Torrey has been a sounding board almost daily throughout my career, and I couldn't ask for a better colleague and friend. At San Diego State University, Matt Lauer has always challenged me to think differently or consider new viewpoints on the endless topics that materialize during our long chats.

A number of colleagues were extraordinarily generous with their time and expertise. In particular, Bob DeLong and Jaret Daniels answered my many questions and educated me on their scientific specialties. They were especially instrumental in helping me think through how and why the past matters in the face of declining biodiversity.

Everyone at Smithsonian Books has been helpful, patient, and supportive. Their vision, hard work, and dedication helped me transform this project into what it is now. Without their efforts, *Understanding Imperiled Earth* would have languished in my mind and on my computer. Julie Huggins and Erika Bűky were instrumental in helping me revise endless drafts. Their feedback was always spot-on and insightful. I am grateful to the entire team at Smithsonian Books for their belief in this project.

Most of all, I am eternally grateful for the love and support of my family. My parents, Craig and Sharon, have always encouraged and supported me through my educational journey and professional career. From an early age, they told me to follow my passions, challenge myself, and take chances. At home, Sopagna and Ellis are my rocks. Through good and bad, they are always there in support. None of this would be possible without them. On a sunny afternoon during our walk home from first grade, Ellis said to me after I shared the good news that this book would be published: "Great, if you want to be rich, you have to write a book." He was right in one regard: my life has been immeasurably enriched by the family, friends, and colleagues who have supported me in my professional endeavors. Thanks to you all!

GLOSSARY OF TERMS

aDNA: DNA extracted from ancient specimens. Studies of aDNA involve recovering genetic material from paleontological, archaeological, and historical materials such as mummified tissues, bone collagen, preserved plant remains, ice cores, marine and lake sediments, and soils to gain information about the evolutionary and genetic histories of plants, animals, and humans.

Agricultural Revolution: A significant transition in human history, when societies shifted from primarily hunter-gatherer lifeways to settled agriculture. This transition occurred independently in different regions of the world, beginning around ten to twelve thousand years ago, and brought about profound changes in human societies and in local and regional ecosystems.

anatomically modern humans (AMHs): A subspecies of the genus *Homo* that includes modern humans, also referred to as *Homo sapiens sapiens*. They are the only surviving members of the *hominin* lineage and the most recent stage in human evolution.

Anthropocene: A proposed geologic epoch, colloquially called the Age of Humans, recognizing the impact of human activities on planet Earth. The current proposals set the beginning of the Anthropocene at roughly fifty years ago, when radionucleotides entered the atmosphere from atomic detonations.

archaeology: The study of the human past from our earliest human ancestors to modern times through the excavation of sites of human activities and the study of artifacts and other physical remains.

biodiversity: The variety and variability of life on Earth, including all living organisms, such as plants, animals, fungi, and microorganisms, along with the ecological systems in which they exist.

Chumash: A Native American cultural group with traditional territories in central and southern California, from Morro Bay in the north to Malibu in the south, including the offshore Northern Channel Islands. Today the Santa Ynez Band of the Chumash Mission Islands tribe is the only federally recognized tribe among more than a dozen bands of Chumash Indians.

climate change: Long-term shifts in weather patterns and average temperatures on Earth, now primarily caused by human activities such as the emission of greenhouse gases.

Clovis: A culture identified by distinctive fluted spear points, once thought to be the earliest peoples to inhabit the Americas. Archaeologists now recognize Clovis peoples as some of the earliest occupants of the Americas but not the first. They lived highly mobile, hunter-gatherer lifestyles between about 13,500 and 12,900 years ago.

conservation biology: An interdisciplinary field focused on protecting and restoring Earth's *biodiversity*, conserving ecosystems, and addressing threats to the natural world.

deep history: The distant past of humans and human societies, reaching back beyond the development of writing systems. The study of deep history integrates perspectives from diverse disciplines, such as geology, archaeology, history, and traditional knowledge.

Early Holocene: A subdivision of the *Holocene* Epoch, typically defined as the time interval from 11,700 to 8,000 years ago.

historical ecology: An interdisciplinary research program that combines history, ecology, and other disciplines to study the long-term interactions between humans and their environments. Through the integration of historical records, archaeological and ecological data, and other sources, historical ecologists reconstruct ancient environments, interpret human interactions with the natural world, and explore the ecological and cultural consequences of these interactions, all with the goal of informing modern conservation efforts.

Holocene: The current geological epoch, typically defined as beginning at 11,700 years ago, often subdivided into the *Early, Middle,* and *Late Holocene* on the basis of climatic shifts.

hominin: A taxonomic group that includes modern humans and our closest extinct relatives, such as members of the genus *Homo* and other bipedal human ancestors.

Industrial Revolution: The socioeconomic and technological transformation that began in Europe in the late eighteenth and early nineteenth centuries, characterized by a shift from an agrarian and labor-intensive economy to one based on industrial production and mechanization.

isotopes: Variants of chemical elements that possess the same number of protons and electrons, but a different number of neutrons in its nucleus. Radioactive isotopes have unstable nuclei that release energy in the form of radiation in order to reach a more stable state. Stable isotopes have stable nuclei that do not emit radiation, decay, or transform to a different state. Archaeologists have used radioactive isotopes such as radiocarbon to date past events and stable isotope analyses to reconstruct diets of ancient humans and other animals.

keystone species: A species that has a disproportionately large impact on the structure and functioning of an ecosystem relative to its abundance or biomass.

land bridge: A connection between two land masses, especially ancient landmasses, that allowed plants, animals, and humans to colonize new territories before the bridge was later submerged by rising sea levels or other geological processes.

Last Glacial Maximum: The most recent period of maximum extent of continental ice sheets and glaciers, roughly twenty thousand years ago. During the Last Glacial Maximum, the Earth's climate was significantly colder than in the present, reshaping the planet's topography, causing drought and desertification, lowering global sea levels, and forcing species to migrate to suitable habitats or face local extinction.

Late Holocene: A subdivision of the *Holocene* Epoch, typically defined as beginning four thousand years ago and extending until the present.

Little Ice Age: A period of cooler than average temperatures that primarily affected the Northern Hemisphere between approximately the fourteenth and the nineteenth centuries. The timing and severity of the Little Ice Age varied across regions, but it resulted in notable impacts on climate, agriculture, economies, and human societies around the world.

mass extinction: A significant and rapid loss of many species on Earth within a relatively short time, resulting in a major decline in

Earth's *biodiversity*. There have been five mass extinction events in Earth's history. Many scientists believe we are living through the sixth mass extinction event.

megafauna: Large-bodied animals whose weights exceed forty-four kilograms (ninety-seven pounds). The term is often used to describe the extinct species from the *Pleistocene* Epoch, such as mammoths, mastodons, saber-tooth cats, giant ground sloths, and giant kangaroos.

midden: A common type of archaeological site consisting of shells, animal bones, botanical remains, artifacts, and other debris that are the refuse of past human activities.

Middle Holocene: A subdivision of the *Holocene* Epoch, typically defined as beginning eight thousand years ago and extending until four thousand years ago.

paleontology: The multidisciplinary study of the fossil and other remains of plants, animals, and other organisms drawing on biology, geology, chemistry, and other disciplines to reconstruct the history and evolution of life on Earth.

palynology: The scientific study of pollen grains, spores, and other microscopic organic particles, combining elements of botany, geology, and ecology to interpret past vegetation patterns, reconstruct ancient climates, investigate human impacts on the environment, and explore the history of plant evolution and landscape change.

Pleistocene: A geological epoch typically defined as beginning 2.6 million years ago and extending until 11,700 years ago. The Pleistocene was the most recent of the Ice Age climatic events, when glaciers covered large parts of the Earth and *megafauna* such as mammoths and mastodons could be found.

169

Quaternary: The most recent period of the Cenozoic Era. The Quaternary includes the *Pleistocene* and *Holocene* Epochs, from 2.6 million years ago to the present day.

radiocarbon dating: A scientific method used to determine the age of organic materials that contain carbon-based compounds, such as animal or plant remains, by measuring the amount of the radioactive isotope carbon-14 (which decays at a constant rate over time) in a sample compared against a reference standard. Radiocarbon dating is the most common technique used by North American archaeologists to determine the age of archaeological sites and artifacts.

restoration: The process of returning an ecosystem that has been degraded, damaged, or destroyed by natural or anthropogenic events to its original or desired condition.

shifting baselines: A concept in conservation biology and ecology that describes a situation where knowledge is gradually lost about the "natural" world because people do not recognize the changes taking place. Shifting baselines can result in a loss of perspective on the extent of human impacts on ecosystems and impede efforts to restore or conserve them.

slash-and-burn: A traditional farming practice used in various parts of the world, particularly in tropical regions with dense vegetation, that involves clearing land by felling vegetation and burning the debris. Crops are then grown in the nutrient-rich ashes.

Southern Dispersal Route: A proposed migratory route followed by early modern humans (see *anatomically modern humans*) out of Africa, following the shores of Africa and the Arabian Peninsula to India, Southeast Asia, and eventually Australia.

RECOMMENDED READING

ARCHAEOLOGY, HISTORY, AND THE MODERN WORLD

Braje, Todd J., Jon M. Erlandson, and Torben C. Rick. *Islands through Time: A Human and Ecological History of California's Northern Channel Islands*. Lanham, MD: Rowman & Littlefield, 2021.

Fagan, Brian M., and Nadia Durrani. *Bigger than History: Why Archaeology Matters*. New York: Thames & Hudson, 2019.

Graeber, David, and David Wengrow. *The Dawn of Everything: A New History of Humanity*. New York: Picador, 2021.

Kelly, Robert L. *The Fifth Beginning: What Six Million Years of Human History Can Tell Us about Our Future*. Oakland: University of California Press, 2019.

Newitz, Annalee. *Four Lost Cities: A Secret History of the Urban Age*. New York: W. W. Norton, 2021.

Shapiro, Beth. *Life as We Made It: How 50,000 Years of Human Innovation Refined—and Redefined—Nature*. New York: Basic Books, 2021.

THE MODERN ENVIRONMENTAL CRISIS

Friedman, Thomas L. *Hot, Flat, and Crowded: Why We Need a Green Revolution—and How We Can Renew Our Global Future*. New York: Farrar, Straus, and Giroux, 2008.

Ghosh, Amitav. *The Great Derangement: Climate Change and the Unthinkable.* Chicago: University of Chicago Press, 2017.

Gilio-Whitaker, Dina. *As Long as Grass Grows: The Indigenous Fight for Environmental Justice, from Colonization to Standing Rock.* Boston: Beacon Press, 2019.

Klein, Naomi. *On Fire: The Burning Case for a Green New Deal.* New York: Simon & Schuster, 2019.

Kolbert, Elizabeth. *The Sixth Extinction: An Unnatural History.* New York: Holt, 2014.

———. *Under a White Sky: The Nature of the Future.* New York: Crown, 2021.

Wilson, E. O. *Half-Earth: Our Planet's Fight for Life.* New York: Liveright, 2017.

Wallace-Wells, David. *The Uninhabitable Earth: Life after Warming.* New York: Tim Duggan, 2019.

Williams, Florence. *The Nature Fix: Why Nature Makes Us Happier Healthier, and More Creative.* New York: W. W. Norton, 2017.

NOTES

CHAPTER 1

6: "specific genes that changed in humans": Iain Mathieson, Iosif Lazaridis, Nadin Rohland, Swapan Mallick, Nick Patterson, Songül Alpaslan Roodenberg, Eadaoin Harney, et al., "Genome-Wide Patterns of Selection in 230 Ancient Eurasians," *Nature* 528 (2015): 499–503.

10: "introduced from Australia in the 1850s": Jared Farmer, *Trees in Paradise: A California History* (New York: W. W. Norton, 2013).

12: "'shifting baselines syndrome' has become a revolutionary concept in environmental conservation": Daniel Pauly, "Anecdotes and the Shifting Baseline Syndrome of Fisheries," *Trends in Ecology and Evolution* 10, no. 10 (1995): 430.

13: "baseline for evaluating subsequent changes": Pauly, "Anecdotes and the Shifting Baseline Syndrome."

14: "northwestern Hawaiian Islands and Palmyra Atoll": Carl Safina, *Song for the Blue Ocean: Encounters along the World's Coasts and beneath the Seas* (New York: Holt, 1999); Carl Safina, *Eye of the Albatross: Visions of Hope and Survival* (New York: Holt, 2003).

15: "based on current data from fish populations around the world": RAM Legacy Stock Assessment Database, https://www.ramlegacy.org/.

16: "important truths about our modern and future world": Carl Safina, "A Shoreline Remembrance," in *Shifting Baselines: The Past and the Future of Ocean Fisheries*, ed. Jeremy B. C. Jackson, Karen E. Alexander, and Enric Sala, 13–19 (Washington, DC: Island Press, 2011).

17: "longer-term datasets include written, archaeological, paleontological, and genetic data": Heike K. Lotze and Loren McClenachan, "Marine Historical Ecology: Informing the Future by Learning from the Past," in *Marine Community Ecology and Conservation*, ed. Mark Bertness, John Bruno, Brian Silliam, and Jay Stachowicz, 165–200 (Oxford: Sinauer Associates, 2013).

17: "'the sea was thick with turtles so numerous'": Carl Safina, "A Shoreline Remembrance," in *Shifting Baselines: The Past and the Future of Ocean Fisheries*, ed. Jeremy B. C. Jackson, Karen E. Alexander, and Enric Sala, 13–19 (Washington, DC: Island Press, 2011).

18: "reconstruct ecosystems before and after human arrival": Heike K. Lotze and Loren McClenachan, "Marine Historical Ecology: Informing the Future by Learning from the Past," in *Marine Community Ecology and Conservation*, ed. Mark Bertness, John Bruno, Brian Silliam, and Jay Stachowicz, 165–200 (Oxford: Sinauer Associates, 2013).

18: "winners from a long-running sport-fishing competition": Loren McClenachan, "Documenting Loss of Large Trophy Fish from the Florida Keys with Historical Photographs," *Conservation Biology* 23, no. 3 (2009): 636–43.

19: "voyage across the American West from 1804 to 1806": Paul S. Martin and Christine R. Szuter, "War Zones and Game Sinks in Lewis and Clark's West," *Conservation Biology* 13, no. 1 (2001): 36–45.

19: "Indigenous hunter-gatherer populations": M. Kat Anderson, *Tending the Wild: Native American Knowledge and the Management of California's Natural Resources* (Berkeley: University of California Press, 2013); Kent G. Lightfoot, Rob Q. Cuthrell, Chuck J. Striplen, and Mark G. Hylkema, "Rethinking the Study of Landscape Management Practices among Hunter-Gatherers in North America," *American Antiquity* 78, no. 2 (2013): 285–301.

21: "assess the health of modern herring populations": Iain McKechnie, Dana Lepofsky, Madonna L. Moss, Virginia L. Butler, Trevor J. Orchard, Gary Coupland,

NOTES

Fredrick Foster, et al., "Archaeological Data Provide Alternative Hypotheses on Pacific Herring (*Clupea pallasii*) Distribution, Abundance, and Variability," *Proceedings of the National Academy of Sciences* 111, no. 9 (2014): E807–E816.

21: "global environments at different points in time": Lotze and McClenachan, "Marine Historical Ecology."

22: "examined and measured nearly fifty thousand oyster shells": Torben C. Rick, Leslie A. Reeder-Myers, Courtney A. Hofman, Denise Breitburg, Rowan Lockwood, Gregory Henkes, Lisa Kellogg, et al., "Millennial-Scale Sustainability of the Chesapeake Bay Native American Oyster Fishery," *Proceedings of the National Academy of Sciences* 113, no. 23 (2016): 6568–73.

23: "reconstructing the abundances, distributions, densities, and food-web links": Heike K. Lotze, Jon M. Erlandson, Marah J. Hardt, Richard D. Norris, Kaustuv Roy, Tim D. Smith, and Christine R. Whitcraft, "Uncovering the Ocean's Past," in *Shifting Baselines: The Past and Future of Ocean Fisheries*, ed. Jeremy B. C. Jackson, Karen Alexander, and Enric Sala, 137–61 (New York: Island Press, 2011).

24: "drive to the brink of extinction": Safina, "A Shoreline Remembrance."

24: "endangered Miami blue butterfly": Scott P. Carroll and Jenella Loye, "Invasion, Colonization, and Disturbance; Historical Ecology of the Endangered Miami Blue Butterfly," *Journal of Insect Conservation* 10 (2006): 13–27.

24: "an unrealistic and impossible goal": Jeffrey Bolster, Karen E. Alexander, and William B. Leavenworth, "The Historical Abundance of Cod on the Nova Scotian Shelf," in *Shifting Baselines: The Past and Future of Ocean Fisheries*, ed. Jeremy B. C. Jackson, Karen Alexander, and Enric Sala, 79–113 (New York: Island Press, 2011).

25: "willing to sacrifice some degree of precision": Jeremy B. C. Jackson and Karen E. Alexander, "Epilogue: Shifting Baselines for the Future," in *Shifting Baselines: The Past and Future of Ocean Fisheries*, ed. Jeremy B. C. Jackson, Karen Alexander, and Enric Sala, 205–6 (New York: Island Press, 2011).

25: "descriptions and comparisons rather than quantitative data": Bolster, Alexander, and Leavenworth, "The Historical Abundance of Cod."

26: "The term was popularized": Paul J. Crutzen and Eugene F. Stoermer, "The 'Anthropocene,'" *Global Change Newsletter* 41 (2000): 17–18.

28: "ten thousand scientists signed an initiative": William Ripple, Christopher Wolf, Thomas Newsome, Phoebe Barnard, William Moomaw, and Philippe Grandcolas, "Scientists' Warning of a Climate Emergency," *Bioscience* 70, no. 1 (2019): 8–12.

CHAPTER 2

33: "extend back another one hundred thousand years": Aurélien Mounier and Marta Mirazón Lahr, "Deciphering African Late Middle Pleistocene Hominin Diversity and the Origin of Our Species," *Nature Communications* 10 (2019): 3406.

33: "AMHs were creating": Richard G. Klein, *The Human Career: Human Biological and Cultural Origins*, 3rd ed. (Chicago: University of Chicago Press, 2009).

34: "new geographic frontiers, starting around 185,000 years ago": Paul Mellars, "Going East: New Genetic and Archaeological Perspectives on Modern Human Colonization of Eurasia," *Science* 313, no. 5788 (2006): 796–800; Israel Hershkovitz, Gerhard W. Weber, Rolf Quam, Mathieu Duval, Rainer Grün, Leslie Kinsley, Avner Ayalon, et al., "The Earliest Modern Humans outside Africa," *Science* 359, no. 6374 (2018): 456–59.

34: "the Southern Dispersal Route": Simon J. Armitage, Sabah A. Jasim, Anthony E. Marks, Adrian G. Parker, Vitaly I. Usik, and Hans-Peter Uerpmann, "The Southern Route 'Out of Africa': Evidence for an Early Expansion of Modern Humans into Arabia," *Science* 331, no. 6016 (2011): 453–56.

34: "mix of terrestrial and marine flora and fauna": Jon M. Erlandson and Todd J. Braje, "Coasting out of Africa: The Potential of Mangrove Forests and Marine Habitats to Facilitate Human Coastal Expansion via the Southern Dispersal Route," *Quaternary International* 382 (2015): 31–41.

35: "between about eighty thousand and fifty thousand years ago": Jane Balme, Iain Davidson, Jo McDonald, Nicola Stern, and Peter Veth, "Symbolic Behavior and the Peopling of the Southern Arc Route to Australia," *Quaternary International* 202 (2009): 59–68; Chris Stringer, "Coasting out of Africa," *Nature* 405 (2000): 24–26.

35: "nearly one hundred kilometers": Michael I. Bird, Scott A. Condie, Sue O'Connor, Damien O'Grady, Christian Reepmeyer, Sean Ulm, Mojca Zega, et al., "Early Human Settlement of Sahul Was Not an Accident," *Scientific Reports* 9, no. 8220 (2019), https://doi.org/10.1038/s41598-019-42946-9.

37: "diverse subsistence economy": Tom D. Dillehay, *Monte Verde: A Late Pleistocene Settlement in Chile*, vol. 2, *The Archaeological Context and Interpretation* (Washington, DC: Smithsonian Institution Press, 1997).

37: "Genetic studies": Jennifer Raff, *Origin* (New York: Twelve, 2022).

37: "following Pacific coastlines": Todd J. Braje, Tom D. Dillehay, Jon M. Erlandson, Richard G. Klein, and Torben C. Rick, "Finding the First Americans," *Science* 358, no. 6363 (2017): 592–94.

37: "twenty-eight thousand years ago": James F. O'Connell and Jim Allen, "When Did Humans Arrive in Greater Australia and Why Is It Important to Know?" *Evolutionary Anthropology* 6, no. 4 (1998): 132–46.

40: "90 of 150 genera driven to extinction": Paul L. Koch and Anthony D. Barnosky, "Late Quaternary Extinctions: State of the Debate," *Annual Review of Ecology, Evolution, and Systematics* 37 (2006): 215–50.

40: "the earliest human-induced biotic crisis in Earth history": Todd J. Braje and Jon M. Erlandson, "Human Acceleration of Animal and Plant Extinctions: A Late Pleistocene, Holocene, and Anthropocene Continuum," *Anthropocene* 4 (2013): 14–23.

40: "extinct approximately 46,000 years ago": T. F. Flannery and R. G. Roberts, "Late Quaternary Extinctions in Australasia: An Overview," in *Extinctions in Near Time: Causes, Contexts, and Consequences*, ed. R. D. E. MacPhee, 239–55 (New York: Plenum, 1999); Richard G. Roberts, Timothy F. Flannery, Linda K. Ayliffe, Hiroyuki Yoshida, Jon M. Olley, Gavin J. Prideaux, Geoff M. Laslett, et al., "New Ages for the Late Australian Megafauna: Continent-Wide Extinction about 46,000 Years Ago," *Science* 292, no. 5523 (2001): 1888–92; Paul L. Koch and Anthony D. Barnosky, "Late Quaternary Extinctions: State of the Debate," *Annual Review of Ecology, Evolution, and Systematics* 37 (2006): 215–50.

40: "thirty-four genera (72 percent) of large mammals went extinct": John Alroy, "Putting North America's End-Pleistocene Megafaunal Extinction in Context: Large-Scale Analyses of Spatial Patterns, Extinction Rates, and Size Distributions," in *Extinctions in Near Time: Causes, Contexts, and Consequences*, ed. R. D. E. MacPhee, 105–43 (New York: Plenum, 1999); Donald K. Grayson, "Late Pleistocene Extinctions in North America: Taxonomy, Chronology, and Explanations," *Journal of World Prehistory* 5 (1991): 193–232; Donald K. Grayson, "Deciphering North American Pleistocene Extinctions," *Journal of Archaeological Research* 63, no. 2 (2007): 185–212.

41: "best explained by the rapid climate changes": James E. King and Jeffrey J. Saunders, "Environmental Insularity and the Extinction of the American Mastodont," In *Quaternary Extinctions: A Prehistoric Revolution*, ed. Paul S. Martin and Richard G. Klein, 315–59 (Tucson: University of Arizona Press, 1984); R. Norman Owen-Smith, *Megaherbivores: The Influence of Very Large Body Size on Ecology* (Cambridge: Cambridge University Press, 1988)

41: "the largest herbivore predators": Jennifer A. Leonard, Carles Vilà, Kena Fox-Dobbs, Paul L. Koch, Robert K. Wayne, and Blaire Van Valkenburgh, "Megafaunal Extinctions and the Disappearance of a Specialized Wolf Ecomorph," *Current Biology* 17, no. 13 (2007): 1146–50; Blaire Van Valkenburgh and Fritz Hertel, "Tough Times at La Brea: Tooth Breakage in Large Carnivores of the Late Pleistocene," *Science* 261, no 5120 (1993): 456–59.

41: "peopling of the Americas at the tail end of the last Ice Age": Paul S. Martin, *Twilight of the Mammoths: Ice Age Extinctions and the Rewilding of America* (Berkeley: University of California Press, 2005); Paul S. Martin, "The Discovery of America: The First Americans May Have Swept the Western Hemisphere and Decimated Its Fauna within 1000 Years," *Science* 179, no. 4077 (1973): 969–74.

43: "arrival of human hunters was enough": William J. Ripple and Blaire Van Valkenburgh, "Linking Top-Down Forces to the Pleistocene Megafaunal Extinctions," *Bioscience* 60, no. 7 (2010): 516–26.

43: "complex feedback loop": Anthony D. Barnosky, Paul L. Koch, Robert S. Feranec, Scott L. Wing, and Alan B. Shabel, "Assessing the Causes of the Late Pleistocene Extinctions on the Continents," *Science* 306, no. 70 (2004): 70–75; Stephen Wroe, Judith Field, and Donald K. Grayson, "Megafaunal Extinction: Climate, Humans, and Assumptions," *Trends in Ecology and Evolution* 21, no. 2 (2006): 61–62.

43: "linked to human population increases": Barnosky et al., "Assessing the Causes."

43: "thousands of times greater than background rates": Gerardo Ceballos, Andrés García, and Paul R. Ehrlich, "The Sixth Extinction Crisis: Loss of Animal Populations and Species," *Journal of Cosmology* 8 (2010): 1821–31.

44: "domestication of wild plants and animals began in southwest Asia": Bruce D. Smith and Melina A. Zeder, "The Onset of the Anthropocene," *Anthropocene* 4 (2013): 8–13.

44: "moved away from generalized hunting and foraging to specialized and intensive agricultural production": Charles L. Redman, *Human Impact on Ancient Environments* (Tucson: University of Arizona Press, 1999); Melina A. Zeder, Daniel G. Bradley, Eve Emshwiller, and Bruce D. Smith, eds., *Documenting Domestication: New Genetic and Archaeological Paradigms* (Berkeley: University of California Press, 2006).

44–45: "Human activity outstripped natural climate change": Smith and Zeder, "The Onset of the Anthropocene."

45: "Rodents, weeds, dogs, and livestock spread around the globe": Redman, *Human Impact on Ancient Environments*.

45: "'Ain Ghazal": Gary O. Rollefson and Ilse Köhler-Rollefson, "Early Neolithic Exploitation Patterns in the Levant: Cultural Impact on the Environment," *Population and Environment: A Journal of Interdisciplinary Studies* 13, no. 4 (1992): 243–54.

48: "trajectory and scale of human impacts": Patrick V. Kirch, "Hawaii as a Model System for Human Ecodynamics," *American Anthropologist* 109, no. 1 (2007): 8–26.

49: "more than 50 percent of the native birds went extinct": Donald K. Grayson, "The Archaeological Record of Human Impacts on Animal Populations," *Journal of World Prehistory* 15 (2001): 1–68.

50: "rodents played a significant role in the extinction": John R. Flenly, "The Paleoecology of Easter Island, and Its Ecological Disaster," in *Easter Island Studies: Contributions to the History of Rapanui in Memory of William T. Mulloy*, ed. Steven R. Fischer, 27–45 (Oxford: Oxbow, 1993).

50: "Islands serve as important models": Scott M. Fitzpatrick and Jon M. Erlandson, "Island Archaeology, Model Systems, the Anthropocene, and How the Past Informs the Future," *Journal of Island and Coastal Archaeology* 13, no. 2 (2018): 283–99.

51: "not until the twentieth century that cities became homes": Redman, *Human Impact on Ancient Environments*.

51: "city's 'ecological shadow'": Sing C. Chew, *World Ecological Degradation: Accumulation, Urbanization, and Deforestation 3000 B.C.–A.D. 2000* (Walnut Creek, CA: AltaMira Press, 2001).

51: "archaeologists have meticulously reconstructed": Damian Evans and Roland Fletcher, "The Landscape of Angkor Wat Redefined," *Antiquity* 89, no. 348 (2015): 1402–19.

52: "Angkor boasted four massive, manmade water reservoirs": Michael D. Coe, *Angkor and the Khmer Civilization* (New York: Thames & Hudson, 2005).

52–53: "'The people of Angkor changed everything'": Richard Stone, "The End of Angkor," *Science* 311, no. 5766 (2006): 1364–68.

54: "evidence of failed spillways and accumulating silt": Stone, "The End of Angkor."

CHAPTER 3

57: "Earth's ecosystems had a moment to recover": P. F. Rupani, M. Nilashi, R. A. Abumallah, S. Asadi, S. Samad, and S. Wang, "Coronavirus Pandemic (COVID-19) and Its Natural Environmental Impacts," *International Journal of Environmental Science and Technology* 17 (2020): 4655–66.

57: "reduced annual carbon emissions by about 7 percent": Pierre Friedlingstein, Michael O'Sullivan, Matthew W. Jones, Robbie M. Andrew, Judith Hauk, et al., "Global Carbon Budget 2020," *Earth System Science Data* 12 (2020): 3269–340.

58: "Methane, which is less abundant but twenty-eight times more potent": NOAA Global Monitoring Laboratory, https://gml.noaa.gov/ccgg/carbontracker-ch4. Accessed April 1, 2022.

58: "sea levels were nearly eighty feet higher": Doyle Rice, "COVID-19 Hasn't Slowed Global Warming: Earth's Carbon Dioxide Levels Highest in Over 3 Million Years, NOAA Says," *USA Today*, April 7, 2021.

59: "'Human activity is driving climate change'": "Despite Pandemic Shutdowns, Carbon Dioxide and Methane Surged in 2020," NOAA Research News, April 7, 2020, research.noaa.gov.

60: "it's now or never": Sam Meredith, "'It's Now or Never': World's Top Climate Scientists Issue Ultimatum on Critical Temperature Unit," *CNBC*, April 4, 2022.

62: "concentrations of methane also rose": William F. Ruddiman and Jonathan S. Thomson, "The Case for Human Causes of Increased Atmospheric CH_4 over the Last 5000 Years," *Quaternary Science Reviews* 20, no. 18 (2001): 1769–77; William F. Ruddiman, Zhengtang Guo, Xin Zhou, Hanbin Wu, and Yanyan Yu, "Early Rice Farming and Anomalous Methane Trends," *Quaternary Science Reviews* 27, nos. 13–14 (2008): 1291–95; William F. Ruddiman, "How Did Humans First Alter Global Climate?" *Scientific American* 292, no. 3 (2005): 46–53.

62: "combination of the natural loss of carbon-rich vegetation": William F. Ruddiman, "On 'The Holocene CO_2 Rise: Anthropogenic or Natural?'" *Eos, Transactions, American Geophysical Union* 87, no. 35 (2022): 352–53.

63: "human activities kept our planet warmer than natural climatic cycles would have dictated": William F. Ruddiman, "The Anthropogenic Greenhouse Era Began Thousands of Years Ago," *Climatic Change* 61 (2003): 261–93.

65: "forest cover peaked in Europe": N. Roberts, R. M. Fyfe. J. Woodbridge, M.-J. Gaillard, B. A. S. Davis, J. O. Kaplan, L. Marquer, et al., "Europe's Lost Forests: A Pollen-Based Synthesis for the Last 11,000 Years," *Scientific Reports* 8, no. 716 (2018), https://doi.org/10.1038/s41598-017-18646-7.

65: "human activities had transformed the face of the planet" Lucas Stephens, Dorian Fuller, Nicole Boivin, Torben Rick, Nicolas Gauthier, Andrea Kay, Ben Marwick, et al., "Archaeological Assessment Reveals Earth's Early Transformation through Land Use," *Science* 365, no. 6456 (2019): 897–902.

65: "farmers in southern China took to flooding lowlands": Ruddiman and Thomson, "The Case for Human Causes of Increased Atmospheric CH_4"; Ruddiman et al., "Early Rice Farming"; Ruddiman, "How Did Humans First Alter Global Climate?"

66: "found a strong correlation": Ruddiman et al., "Early Rice Farming."

66: "these findings lend support to Ruddiman's conclusions": Dorian Q. Fuller, Jacob van Etten, Katie Manning, Cristina Castillo, Eleanor Kingwell-Banham, Alison Weisskopf, Ling Qi, et al., "The Contribution of Rice Agriculture and Livestock Pastoralism to Prehistoric Methane Levels: An Archaeological Assessment," *Holocene* 21, no. 5 (2001): 743–59.

66: "transform steep hillsides in Southeast Asia into rice paddies": Ruddiman, "How Did Humans First Alter Global Climate?"

66: "increases of 40 ppm in atmospheric carbon dioxide and 250 ppb for methane": Ruddiman, "How Did Humans First Alter Global Climate?"

67: "sixteenth-century Icelandic poem": Tamie J. Jovanelly, *Iceland: Tectonics, Volcanics, and Glacial Features* (Hoboken, NJ: American Geophysical Union, 2020).

68: "creating a feedback cycle": Gifford H. Miller, Áslaug Geirsdóttir, Yafang Zhong, Darren J. Larsen, Bette L. Otto-Bliesner, Marika M. Holland, David A. Bailey, et al., "Abrupt Onset of the Little Ice Age Triggered by Volcanism and Sustained by

Sea-Ice/Ocean Feedback," *Geophysical Research Letters* 39, no. 2 (2012), https://doi .org/10.1029/2011GL050168.

69: "60.5 million people living in the Americas in 1492": Alexander Koch, Chris Brierley, Mark M. Maslin, and Simon L. Lewis, "Earth System Impacts of the European Arrival and Great Dying in the Americas after 1492," *Quaternary Science Reviews* 207 (2019): 13–36.

70: "fifty-five million Native American deaths by 1600": Koch et al., "Earth System Impacts."

70–71: "Fifty-six million hectares (two hundred thousand square miles) of agricultural fields reverted to forest during the sixteenth century": Koch et al., "Earth System Impacts."

72: "impacts of forest and animal-grazing management on global vegetation": Karl-Heinz Erb, Thomas Kastner, Christoph Plutzar, Anna Liza S. Bais, Nuno Caralhais, Tamara Fetzel, Simone Gingrich, et al., "Unexpectedly Large Impact of Forest Management and Grazing on Global Vegetation Biomass," *Nature* 553 (2018): 73–76.

CHAPTER 4

74: "At 6:20 p.m. on April 14, 2019": "What We Know and Don't Know about the Notre-Dame Fire," *New York Times*, April 17, 2019.

75: "Notre-Dame's wood and metal roof was destroyed": "France Vows to Rebuild Notre Dame Cathedral after Devastating Fire," *CBS News*, April 17, 2019.

75: "more than a billion dollars had been offered": "Notre-Dame Fire: Millions Pledged to Rebuild Cathedral," *BBC News*, April 16, 2019.

75: "'There are no longer trees of that size in France'": "There Are No Trees in France That Are Big Enough to Rebuild Notre Dame's Roof," *CNN World*, April 16, 2019.

76: "continuous human occupations by agriculturalists likely triggered successive cycles of deforestation": Jed O. Kaplan, Kristen M. Krumhardt, and Niklaus Zimmermann, "The Prehistoric and Preindustrial Deforestation of Europe," *Quaternary Science Reviews* 28 (2009): 3016–34.

76: "'single factor that has changed the European landscape'": Michael Williams, "Dark Age and Dark Areas: Global Forestation in the Deep Past," *Journal of Historical Geography* 26, no. 1 (2000): 28–46.

76: "northern reaches of the continent were blanketed in an ice sheet": Neil Roberts, *The Holocene: An Environmental History*, 2nd ed. (Oxford: Blackwell, 1998).

76: "woolly mammoths standing 3.4 meters (11 feet) tall": Ross D. E. MacPhee, *End of the Megafauna: The Fate of the World's Hugest, Fiercest, and Strangest Animals* (New York: W. W. Norton, 2018).

77: "a return to Ice Age conditions": Hans Renssen, Aurélien Mairesse, Hugues Goosse, Pierre Mathiot, Oliver Heiri, Didier M. Roche, Kerim H. Nisancioglu, and Paul J. Valdes, "Multiple Causes of the Younger Dryas Cold Period," *Nature Geoscience* 8 (2015): 946–49.

77: "deciduous trees such as oak, elm, and beech took their place": Neil Roberts, *The Holocene: An Environmental History*, 2nd ed. (Oxford: Blackwell, 1998).

78: "led to large permanent villages": T. Douglas Price and Anne Birgitte Gebauer, eds., *Last Hunters—First Farmers: New Perspectives on the Prehistoric Transition to Agriculture* (Santa Fe, NM: School of American Research, 1996).

79: "they hunted and gathered boar, deer, gazelle, fish, and mollusks": Ofer Bar-Yosef, "The Natufian Culture of the Levant, Threshold to the Origins of Agriculture," *Evolutionary Anthropology* 6, no. 5 (1998): 159–77.

79: "Wild cereals (plants that produce starchy grains)": Dolores R. Piperno, Ehud Weiss, Irene Holst, and Dani Nadel, "Processing of Wild Cereal Grains in the Upper Paleolithic Revealed by Starch Grain Analysis," *Nature* 407 (2004): 607–73.

79: "people began selecting, planting, and tending large, hardy seeds": Ofer Bar-Yosef, "The Natufian Culture of the Levant, Threshold to the Origins of Agriculture," *Evolutionary Anthropology* 6, no. 5 (1998): 159–77.

79: "legumes such as lentils and peas": Naomi F. Miller, "The Origins of Plant Cultivation in the Near East," in *The Origins of Agriculture: An International Perspective*, ed. C. Wesley Cowan and Patty Jo Watson, 39–58 (Tuscaloosa: University of Alabama Press, 2006).

79: "experimenting with the domestication of animals": Melinda A. Zeder and Brian Hesse, "The Initial Domestication of Goats (*Capra hircus*) in the Zagros Mountains 10,000 Years Ago," *Science* 287 (2000): 2254–57; Gordon Luikart, Ludovic Gielly, Laurent Excoffier, Jean-Denis Vigne, Jean Bouvet, and Pierre Taberlet,

"Multiple Maternal Origins and Weak Phylogeographic Structure in Domestic Goats," *Proceedings of the National Academy of Sciences* 98, no. 10 (2001): 5927–32; Mary Stiner, C. Hijlke Buitenhuis, Güneş Duru, Steven L. Kuhn, Susan M. Mentzer, Natalie D. Munro, Nadja Pöllath, Jay Quade, Georgia Tsartsidou, and Mihriban Özbaşaran, "A Forager–Herder Trade-Off, from Broad-Spectrum Hunting to Sheep Management at Aşıklı Höyük, Turkey," *Proceedings of the National Academy of Sciences* 111, no. 23 (2014): 8404–9; Ruth Bollongino, Joachim Burger, Adam Powell, Marjan Mashkour, Jean-Denis Vigne, and Mark G. Thomas, "Modern Taurine Cattle Descended from Small Number of Near-Eastern Founders," *Molecular Biology and Evolution* 29, no. 9 (2012): 2101–4.

80: "exclusive reliance on sheep": Stiner et al., "A Forager–Herder Trade-Off."

80: "Genetic studies suggest that the mechanism was likely migration": Guido Brandt Szécsényi-Nagy, Wolfgang Haak, Victoria Keerl, János Jakucs, Sabine Möller-Rieker, Kitti Köhler, Balázs Gusztáv Mende, Krisztián Oross, Tibor Marton, et al., "Tracing the Genetic Origin of Europe's First Farmers Reveals Insights into Their Social Organization," *Proceedings of the Royal Society B* 282 (2015), https://doi .org/10.1098/rspb.2015.0339.

81: "established by migrants from the Near East and Anatolia": Wolfgang Haak, Oleg Balanovsky, Juan J. Sanchez, Sergey Koshel, Valery Zaporozhchenko, Christina J. Adler, Clio S. I. der Sarkissian, Guido Brandt, Carolin Schwarz, Nicole Nicklisch, et al., "Ancient DNA from European early Neolithic Farmers Reveals Their Near Eastern Affinities," *PLoS Biology* 2010, https://doi.org/10.1371/journal .pbio.1000536.

81: "arrived in European landscapes about nine thousand years ago": Iosif Lazardis, Nick Patterson, Alissa Mittnik, Gabriel Renaud, Swapan Mallick, Karola Kirsanow, Peter H. Sudmant, Joshua G. Schraiber, Sergi Castellano, Mark Lipson, et al., "Ancient Human Genomes Suggest Three Ancestral Populations for Present-Day Europeans," *Nature* 513 (2014): 409–13.

81: "Genetic analysis of ancient and modern populations": Zuzana Hofmanová, Susanne Kreutze, Garrett Hellenthal, Christian Sell, Yoan Diekmann, David Díez-del-Molino, Lucy van Dorp, Saioa López, Athanasios Kousathanas, Vivian Link, et al., "Early Farmers from across Europe Directly Descended from Neolithic Aegeans," *Proceedings of the National Academy of Sciences* 113, no. 25 (2015): 6886–91.

NOTES

81: "spread rapidly across central Europe": Corrie Bakels, "The First Farmers of the Northwest European Plain: Some Remarks on Their Crops, Crop Cultivation and Impact on the Environment," *Journal of Archaeological Science* 51 (2014): 94–97.

83: "dwellings were 5.5–7 meters (18–23 feet) wide": Arkadiusz Marciniak, *Placing Animals in the Neolithic: Social Zooarchaeology of Prehistoric Farming Communities* (London: UCL Press, 2005).

83: "slowly absorbed into agricultural communities": Alex Bentley, "Mobility, Specialisation and Community Diversity in the Linearbandkeramik: Isotopic Evidence from the Skeletons," *Proceedings of the British Academy* 144 (2007): 117–140.

83: "Natural landscapes became anthropogenic ones": Bakels, "The First Farmers."

83: "clearance of forests and woodlands": H. C. Darby, "The Clearing of the Woodland in Europe," In *Man's Role in Changing the Face of the Earth*, ed. William L. Thomas Jr., 183–216 (Chicago: University of Chicago Press, 1956).

85: "anthromes": Erle C. Ellis and Navin Ramankutty, "Putting People in the Map: Anthropogenic Biomes of the World," *Frontiers in Ecology and the Environment* 6 (2008): 439–47.

85: "data collection began only in the mid-1970s": Matthew C. Hansen and Thomas R. Loveland, "A Review of Large Area Monitoring of Land Cover Change Using Landsat Data," *Remote Sensing of Environment* 122 (2012): 66–74.

86: "European forest cover over the past ten thousand years": N. Roberts, R. M. Fyfre, J. Woodbridge, M.-J. Giallard, B. A. S. Davis, J. O. Kaplan, L. Marquer, et al., "Europe's Lost Forest: A Pollen-Based Synthesis for the Last 11,000 Years," *Scientific Reports* 8 (2018): 716.

87: "burning of forests by hunter-gatherer groups or natural climatic changes": James Innes, Jeffrey Blackford, and Ian Simmons, "Woodland Disturbance and Possible Land-Use Regimes during the Late Mesolithic in the English Uplands: Pollen, Charcoal and Non-pollen Palynomorph Evidence from Bluewath Beck, North York Moors, UK," *Vegetation History and Archaeobotany* 19, nos. 5–6 (2010): 439–52.

87: "arrival of agriculturalists corresponds with an increase in secondary woodland": Jutta Lechterbeck, Kevan Edinborough, Tim Kerig, Ralph Fyfe, Neil Roberts, and Stephen Shennan," Is Neolithic Land Use Correlated with Demography? An

Evaluation of Pollen-Derived Land Cover and Radiocarbon-Inferred Demographic Change from Central Europe," *Holocene* 24, no. 10 (2014): 1297–1307.

87: "role of humans in the transformation of European landscapes": Michael Williams, "Dark Ages and Dark Areas: Global Deforestation in the Deep Past," *Journal of Historical Geography* 26, no. 1 (2000): 28–46.

88: "Historical writings detail the loss of European forests": Williams, "Dark Ages and Dark Areas."

88: "grazing by domesticated cattle, pigs, and sheep helped make deforestation permanent": Sing C. Chew, *World Ecological Degradation: Accumulation, Urbanization, and Deforestation 3000 B.C.–A.D. 2000* (Walnut Creek, CA: AltaMira Press, 2001).

88: "classical texts note warmer temperatures": Chew, *World Ecological Degradation.*

88: "pagan beliefs and local religious traditions had bestowed special significance": Charles L. Redman, *Human Impact on Ancient Environments* (Tucson: University of Arizona Press, 1999).

89: "privileges humans above all other forms of creation": Daniel Hillel, *Out of Earth: Civilization and the Life of the Soil* (Berkeley: University of California Press, 1992).

89: "regrowth of secondary forests in many regions of Europe": Roberts et al., "Europe's Lost Forest."

89: "Populations surged over the next seven hundred years": Williams, "Dark Ages and Dark Areas."

89: "transformed into productive agricultural lands by forest clearance": Archibald R. Lewis, "The Closing of the Medieval Frontier, 1250–1350," *Speculum* 33, no. 4 (1958): 465–83.

89–90: "enhanced people's ability to cultivate difficult terrain": Lynn White Jr., *Medieval Technology and Social Change* (London: Oxford University Press, 1964).

90: "more than half that area had been converted": Michael Williams, *Deforesting the Earth: From Prehistory to Global Crisis; An Abridgment* (Chicago: University of Chicago Press, 2006).

90: "This document is unique": Oliver Rackham, *Ancient Woodland: Its History, Vegetation and Uses in England*, 2nd ed. (Colvend, UK: Castlepoint, 2003).

90: "wiped out at least one-third of the population": Williams, *Deforesting the Earth.*

92: "supply of timber diminished and prices rose": Williams, *Deforesting the Earth.*

92: "only 5 to 10 percent of Europe remained forested": Chew, *World Ecological Degradation.*

94: "the French government decided to rebuild the roof and spire as replicas": "A Look at Notre-Dame Cathedral's Restoration Two Years after Devastating Fire," Afar .com, April 15, 2021.

95: "humans have reduced the global number of trees": T. W. Crowther, H. B. Glick, K. R. Covey, C. Bettigole, D. S. Maynard, S. M. Thomas, J. R. Smith, et al., "Mapping Tree Density at a Global Scale," *Nature* 525 (2015): 201–5.

96: "access to green space and trees": Sandra Bogar and Kristen M. Beyer, "Green Space, Violence, and Crime: A Systematic Review, "*Trauma, Violence, and Abuse* 17, no. 2 (2016): 160–71.

96: "committed to end deforestation by 2030": "Will the COP26 Global Deforestation Pledge Save Forests? "*National Geographic,* November 4, 2021.

CHAPTER 5

101: "142 distinct bird species found nowhere else on the planet": Stefan Verbano, "Hawai'i's Endangered Birds," *Ke Ola: Hawai'i Island's Community Magazine,* September–October 2019.

102: "One-third of the endangered bird species": Brooke Jarvis, "'How Scientists Are Racing to Save a Rare Hawaiian Bird from Extinction," *Audubon,* September–October 2015.

102: "the black rat (*Rattus rattus*)": Jesse Geenspan, "Why Hawaii Is the Epicenter of the Avian Extinction Crisis," *BirdWatching,* December, 3 2021.

103: "avian malaria starting sometime in the early twentieth century": Carter T. Atkinson and Dennis A. LaPointe, "Introduced Avian Diseases, Climate Change, and the Future of Hawaiian Honeycreepers," *Journal of Avian Medicine and Surgery* 23, no. 1 (2009): 53–63.

103: "150 species are driven to extinction every single day": Fred Pearce, "Global Extinction Rates: Why Do Estimates Vary So Wildly?" *YaleEnvironment360,* August 17, 2015.

104: "later pulses were likely triggered": David M. Raup and J. John Sepkoski Jr., "Mass Extinctions in the Marine Fossil Record," *Science* 215, no. 4539 (1982): 1501–3.

104: "Also wiped out were": Paul R. Renne, Alan L. Deino, Frederik J. Hilgen, Klaudia F. Kuiper, Darren F. Mark, William S. Michell III, Leah E. Morgan, et al., "Time Scales of Critical Events around the Cretaceous-Paleogene Boundary," *Science* 339, no. 6120 (2013): 684–87.

104: "About 98 percent of all the plant and animal species": Tammana Begum, "What Is Mass Extinction and Are We Facing a Sixth One?" *Natural History Museum*, May 19, 2021.

105: "one hundred to one thousand times higher than background levels": Peter M. Vitousek, Harold A. Mooney, Jane Lubchenco, and Jerry M. Melillo, "Human Domination of Earth's Ecosystems," *Science* 277, no. 5325 (1997): 494–9.

106: "Humans have already transformed": Begum, "What Is Mass Extinction?"

107: "face increased likelihood of extinction": International Union for the Conservation of Nature, "Species and Climate Change," *Issues Brief*, December 2019.

108: "modern land- and seascapes": Todd J. Braje and Jon M. Erlandson, "Human Acceleration of Animal and Plant Extinctions: A Late Pleistocene, Holocene, and Anthropocene Continuum," *Anthropocene* 4 (2013): 14–23.

109: "Miami blues could be found": J. V. Calhoun, J. R. Slotten, and M. H. Salvato, "The Rise and Fall of Tropical Blues in Florida: *Cyclargus ammon* and *Cyclargus thomasi bethunebakeri*," *Holartic Lepidoptera* 7, no. 1 (2002): 13–20.

109: "in only a few locations along the Florida Keys": Scott P. Carroll and Jenella Loye, "Invasion, Colonization, and Disturbance; Historical Ecology of the Endangered Miami Blue Butterfly," *Journal of Insect Conservation* 10 (2006): 13–27.

110: "Miami blue butterfly was granted federal endangered-species status": Calhoun, Slotten, and Salvato, "The Rise and Fall of Tropical Blues."

111: "butterflies were commonly found in association with balloon vine": Carroll and Loye, "Invasion, Colonization, and Disturbance."

113: "repair Earth's damaged ecosystems": Holly P. Jones, Peter C. Jones, Edward B. Barbier, Ryan C. Blackburn, Jose M. Rey Benayas, Karen D. Holl, Michelle McCrackin, et al., "Restoration and Repair of Earth's Damaged Ecosystems,"

Proceedings of the Royal Society B 285, no. 1873 (2018), https://doi.org/10.1098/rspb
.2017.2577.

113: "The famous entomologist E. O. Wilson": Jeremy Hance, "'Could We Set Aside
Half the Earth for Nature?'" *Guardian,* June 15, 2016.

113: "seal and sea lion populations were driven to near extinction": Charles M.
Scammon, *The Marine Mammals of the North-Western Coast of North America,
Described and Illustrated: Together with an Account of the American Whale-Fishery*
(San Francisco: John H. Carmony, 1874).

114: "No Guadalupe fur seals were sighted": Richard Ellis, *The Empty Ocean*
(Washington, DC: Island Press, 2003).

114: "discovered in a cave on Isla Guadalupe": Marianne Riedman, *The Pinnipeds:
Seals, Sea Lions, and Walruses* (Berkeley: University of California Press, 1990).

114–15: "Guadalupe fur seals are listed as a threatened species": Ellis, *The Empty
Ocean.*

115: "commercial hunting devastated whale populations": Scammon, *The Marine
Mammals of the North-Western Coast.*

116: "oldest remains are at least eleven thousand years old": Torben C. Rick,
Robert L. DeLong, Jon M. Erlandson, Todd J. Braje, Terry L. Jones, Douglas J.
Kennett, Thomas A. Wake, and Phillip L. Walker. "A Trans-Holocene Archaeological
Record of Guadalupe Fur Seals (*Arctocephalus townsendi*) on the California Coast,"
Marine Mammal Science 25, no. 2 (2009.): 487–502.

116: "date back to about seven thousand years ago": Torben C. Rick, Robert L.
DeLong, Jon M. Erlandson, Todd J. Braje, Terry L. Jones, Jeanne E. Arnold,
Matthew R. DesLauriers, et al., "Where Were the Northern Elephant Seals?
Holocene Archaeology and Biogeography of *Mirounga angustirostris*," *Holocene* 21,
no. 7 (2011): 1159–66.

116: "archaeological deposit approximately twelve thousand years old": Courtney A.
Hofman, Torben C. Rick, Jon M. Erlandson, Leslie Reeder-Myers, Andreanna J.
Welch, and Michael Buckley, "Collagen Fingerprinting and the Earliest Marine
Mammal Hunting in North America," *Scientific Reports* 8 (2018): 10014.

121: "viruses have passed from animal hosts to humans": Andrew P. Dobson,
Stuart L. Pimm, Lee Hannah, Les Kaufman, Jorge A. Ahumada, Amy W. Ando, Aaron

Bernstein, et al., "Ecology and Economics for Pandemic Prevention," *Science* 369, no. 6502 (2020): 379–81.

122: "5.6 gigatons of carbon": Sandra Myrna Díaz, Josef Settele, Eduardo Brondízio, Hien Ngo, Maximilien Guéze, John Agard, Almut Arneth, et al., *The Global Assessment Report on Biodiversity and Ecosystem Services: Summary for Policy Makers.* IPBES, 2019.

123: "Indigenous peoples who occupy a quarter of the Earth's land surface": Diaz et al., *Global Assessment Report.*

CHAPTER 6

126: "sell for up to $40 per pound": "Bluefin Goes for $3 Million at 1st 2019 Sale at Tokyo Market," Associated Press, January 5, 2019.

128: "depleted worldwide fish stocks": U. Rashid Sumaila and Travis C. Tai, "End Overfishing and Increase the Resilience of the Ocean to Climate Change," *Frontiers in Marine Science* 7 (2020), https://doi.org/10.3389/fmars.2020.00523.

128: "between 40 and 70 percent of fish stocks are now estimated to be at unsustainable levels": Sumaila and Tai, "End Overfishing."

128: "'fishing down marine food webs'": Daniel Pauly, Villy Christensen, Johanne Dalsgaard, Rainer Froese, and Francisco Torres, Jr., "Fishing down Marine Food Webs," *Science* 279, no. 5352 (1998): 860–63.

129: "harvesting the Grand Banks of the western Atlantic": Poul Holm, Francis Ludlow, Cordula Scherer, Charles Travis, Bernard Allaire, Cristina Brito, Patrick W. Haye, et al., "The North Atlantic Fish Revolution (ca. AD 1500)," *Quaternary Research* 108 (2019), https://doi.org/10.1017/qua.2018.153; David C. Orton, James Morris, Alison Locker, and James H. Barrett, "Fish for the City: Meta-analysis of Archaeological Cod Remains and the Growth of London's Northern Trade," *Antiquity* 88, no. 340 (2014): 516–30.

129: "stock sizes were diminishing": Peter Jones, Alison Cathcart, and Douglas C. Speirs, "Early Evidence of the Impact of Preindustrial Fishing on Fish Stocks from the Mid-west and Southeast Coastal Fisheries of Scotland in the 19th Century," *ICES Journal of Marine Science* 73, no. 5 (2016): 1404–14.

NOTES

130: "home to animal and plant species found nowhere else on Earth": Todd J. Braje, Jon M. Erlandson, and Torben C. Rick, *Islands through Time: A Human and Ecological History of California's Northern Channel Islands* (Lanham, MD: Rowman & Littlefield, 2021).

132: "adjust their subsistence strategies as their communities grew": Braje, Erlandson, and Rick, *Islands through Time.*

132: "archaeological records of the Island Chumash": Jeanne E. Arnold, ed. *The Origins of a Pacific Coast Chiefdom: The Chumash of the Channel Islands* (Salt Lake City: University of Utah Press, 2001).

132: "millennia-long perspective": Torben C. Rick, *The Archaeology and Historical Ecology of Late Holocene San Miguel Island* (Los Angeles, CA: Cotsen Institute of Archaeology, 2007).

133: "shellfish provided about 90 percent of the animal protein": Jon M. Erlandson, Torben C. Rick, and Todd J. Braje, "Fishing up the Food Web? 12,000 Years of Maritime Subsistence and Adaptive Adjustments on California's Channel Islands," *Pacific Science* 63, no. 4 (2009): 711–24.

134: "kelp-forest fish contributed the bulk of dietary protein": Torben C. Rick, Jon M. Erlandson, and René L. Vellanoweth, "Paleocoastal Marine Fishing on the Pacific Coast of the Americas: Perspectives from Daisy Cave, California," *American Antiquity* 66, no. 4 (2001): 595–614.

135: "that at many Middle Holocene sites": Torben C. Rick, René L. Vellanoweth, Jon M. Erlandson, and Douglas J. Kennett, "On the Antiquity of the Single-Piece Shell Fishhook: AMS Radiocarbon Evidence from the Southern California Coast," *Journal of Archaeological Science* 29, no. 9 (2002): 933–42.

135: "focused on low- to mid-trophic-level species": Erlandson, Rick, and Braje, "Fishing up the Food Web?"

135: "To facilitate fishing and sea-mammal hunting": Braje, Erlandson, and Rick, *Islands through Time.*

137: "northern site was the epicenter of village life after Spanish contact": Todd J. Braje, Torben C. Rick, Leslie Reeder-Myers, Breana Campbell, and Kelly Minas, "Defining the Historic Landscape on Eastern Santa Rosa Island: Archeological

Investigations at *Qshiwqshiw*," *Monographs of the Western North American Naturalist* 7 (2014): 135–45.

138: "small-scale excavations of the southern locus of Qshiwqshiw": Braje et al., "Defining the Historic Landscape."

139: "propelled fishers to focus more on fish": Jon M. Erlandson, Torben C. Rick, Todd J. Braje, Alexis Steinberg, and René Vellanoweth, "Human Impacts on Ancient Shellfish: A 10,000 Year Record from San Miguel Island, California," *Journal of Archaeological Science* 35, no. 8 (2008): 2144–52.

139: "The most common fish captured at Qshiwqshiw were surfperches": Emma A. Elliott Smith, Todd J. Braje, Kenneth W. Gobalet, Seth D. Newsome, Breana Campbell, and Torben C. Rick, "Archaeological and Stable Isotope Data Reveal Patterns of Fishing Across the Food Web on California's Channel Islands," *The Holocene* 33, no. 4 (2023): 446–58.

141: "establishing fishing baselines built from deep historical data": Jeremy B. C. Jackson, Michael X. Kirby, Wolfgang H. Berger, Karen A. Bjorndal, Louis W. Botsford, Bruce J. Bourque, et al., "Historical Overfishing and the Recent Collapse of Coastal Ecosystems," *Science* 293, no. 5530 (2001): 629–37.

141: "effective in combating the shifting baselines syndrome": Arnault Le Bris, Andrew J. Pershing, Christina M. Hernandez, Katherine E. Mills, and Graham D. Sherwood, "Modelling the Effects of Variation in Reproductive Traits on Fish Population Resilience," *ICES Journal of Marine Science* 72, no. 9 (2015): 2590–99.

141: "tend to produce more offspring": Boris Worm, Ray Hilborn, Julia K. Baum, Trevor A. Branch, Jeremy S. Collie, Christopher Costello, et al. "Rebuilding Global Fisheries, " *Science* 325, no. 5940 (2009): 578–85; Daniel C. Gwinn, Micheal S. Allen, Fiona D. Johnston, Paul Brown, Charles R. Todd, and Robert Arlinghaus, "Rethinking Length-Based Fishing Regulations: The Value of Protecting Old and Large Fish with Harvest Slots," *Fish and Fisheries* 16, no. 2 (2015): 259–81.

142: "used fish-bone remains from Chumash archaeological sites": Todd J. Braje, Torben C. Rick, and Jon M. Erlandson, "Rockfish in the Long View: Applied Zooarchaeology and Conservation of Pacific Red Snapper (Genus *Sebastes*) in Southern California," in *Conservation Biology and Applied Zooarchaeology*, ed. Steve Wolverton and R. Lee Lyman, 157–78 (Tucson: University of Arizona Press, 2012); Todd J. Braje, Torben C. Rick, Paul Szpak, Seth D. Newsome, Joseph M. McCain,

NOTES

Emma A. Elliott Smith, et al., "Historical Ecology and the Conservation of Large, Hermaphroditic Fishes in Pacific Coast Kelp Forest Ecosystems," *Science Advances* 3, no. 2 (2017), https://www.doi.org/10.1126/sciadv.1601759.

142: "commercial harvesting of rockfish": Milton S. Love, Mary Yoklavich, and Lyman Thorsteinson, *The Rockfishes of the Northeast Pacific* (Berkeley: University of California Press, 2002).

142: "widespread closures": Braje et al., "Rockfish in the Long View."

144: "They are hermaphrodites": Milton S. Love, Catherine W. Mecklenburg, T. Anthony Mecklenburg, and Lyman K. Thorsteinson, *Resource Inventory of Marine and Estuarine Fishes of the West Coast and Alaska: A Checklist of North Pacific and Arctic Ocean Species from Baja California to the Alaska-Yukon Border* (Washington, DC: US Geological Survey, Biological Resources Division, 2005).

144: "smaller than the average rockfish captured by the Chumash": Braje et al., "Rockfish in the Long View."

146: "dramatic toll on abalone populations": L. Ignacio Vilchis, Mia J. Tegner, James D. Moore, Carolyn S. Friedman, Kristin L. Riser, Thea T. Robbins, and Paul K. Dayton, "Ocean Warming Effects on Growth, Reproduction, and Survivorship of Southern California Abalone," *Ecological Applications* 15, no. 2 (2005): 469–80.

147: "Red abalone was the last of California's abalone species": Todd J. Braje, *Shellfish for the Celestial Empire: The Rise and Fall of Commercial Abalone Fishing in California* (Salt Lake City: University of Utah Press, 2016).

148: "the availability of red abalone was discontinuous": Todd J. Braje, Jon M. Erlandson, Torben C. Rick, Paul K. Dayton, and Marco B. A. Hatch, "Fishing from Past to Present: Continuity and Resilience of Red Abalone Fisheries on the Channel Islands, California," *Ecological Applications* 19, no. 4 (2009): 906–19.

149: "sea surface temperatures were colder": Braje et al., "Fishing from Past to Present."

150: "50 percent of the Earth's surface consists of ocean waters": "What Happens If We Don't Protect the High Seas?" Nature Conservancy, October 16, 2018, www.nature.org/en-us/what-we-do/our-insights/perspectives/what-happens-if-we-dont-protect-the-high-seas.

151: "Fish provide 17 percent of all the animal protein consumed globally": "What Happens If We Don't Protect the High Seas?"

151: "Fish are the single most traded food commodity on Earth": "What Happens If We Don't Protect the High Seas?"

EPILOGUE

155: "34 billion tons of atmospheric carbon dioxide": Peter Brannen, "The Anthropocene Is a Joke," *The Atlantic,* August 13, 2019.

156: "concept of the Anthropocene": Paul J. Crutzen, "Geology of Mankind," *Nature* 415, no. 23 (2002), 23 (2002), https://doi.org/10.1038/415023a.

156: "conversations about the state of the world": Todd J. Braje and Matthew Lauer, "A Meaningful Anthropocene? Golden Spikes, Transitions, Boundary Objects, and Anthropogenic Seascapes," *Sustainability* 12, no. 16 (2020): 6459.

158: "requirement can be found in legislation": Julie Lurman Joly, Joel Reynolds, and Martin Robards, "Recognizing When the 'Best Scientific Data Available' Isn't," *Stanford Environmental Law Journal* 29 (2010): 247–82.

159: "advocates for what he calls 'action archaeology'": Jeremy A. Sabloff, *Archaeology Matters: Action Archaeology in the Modern World* (Walnut Creek, CA: Left Coast Press, 2008).

159–60: "scientists believe that its application to ecological restoration is impractical": Marleen Buizer, Tim Kurz, and Katinka Ruthrof, "Understanding Restoration Volunteering in a Context of Environmental Change: In Pursuit of Novel Ecosystems or Historical Analogues?" *Human Ecology* 40 (2012): 153–60; Young D. Choi, "Restoration Ecology to the Future: A Call for New Paradigm," *Restoration Ecology* 15, no. 2 (2007): 351–53.

INDEX

Note: Illustrations are indicated by page numbers in italics.

INDEX

INDEX